Science of Science

Science of Science

Understanding the Foundations and Limits of Science from an Interdisciplinary Perspective

ALEXANDER KRAUSS

OXFORD
UNIVERSITY PRESS

OXFORD
UNIVERSITY PRESS

Great Clarendon Street, Oxford, OX2 6DP,
United Kingdom

Oxford University Press is a department of the University of Oxford.
It furthers the University's objective of excellence in research, scholarship,
and education by publishing worldwide. Oxford is a registered trade mark of
Oxford University Press in the UK and in certain other countries

Published in the United States of America by Oxford University Press
198 Madison Avenue, New York, NY 10016, United States of America

British Library Cataloguing in Publication Data
Data available

Library of Congress Control Number is on file with the LOC.

ISBN 9780198937371

DOI: 10.1093/9780198937401.001.0001

Printed and bound by
CPI Group (UK) Ltd, Croydon, CR0 4YY

Contents

Introduction 1
1. Describing Science of Science 16
2. Biology of Science 23
3. Archaeology of Science 27
4. Cognitive Science of Science 33
5. Psychology of Science 38
6. Sociology of Science 41
7. Economics of Science 48
8. History of Science 53
9. Anthropology of Science 60
10. Methodology of Science 66
11. Scientometrics and Network Science 74
12. Computer Science of Science 79
13. Statistics and Mathematics of Science 83
14. Philosophy of Science 91
15. Linguistics of Science 101
16. Science of Science: An Integrated and Methods-Driven Understanding of Science 104
17. Science of Science: An Integrated Field Grounded in the New-Methods-Drive-Science Theory 114
18. The Limits of Science: An Overview 118
19. The Limits of Science: Grounded in the Boundaries of Our Methods and Mind 128
20. The Limits of Science: Expanding the Limits by Expanding Our Methodological Toolbox 144

Conclusion and Implications 161

Appendix 169

References 170
Acknowledgements 184
Index 185

Introduction

Science has driven our remarkable advances in modern society. But we do not yet understand well some of the most fundamental questions about science: What are the origins, foundations and boundaries of science? How have we learned what we know about the world around us? Why is it that how we advance science is poorly understood, even though it has an enormous influence on our lives through the medicine we take and the technology we use? We go through our daily lives understanding little about the answers to these fundamental questions about science and the present limits of what we know about the world. Yet these are among the most important questions about science and understanding the world around us. No consensus yet exists on these fundamental questions, as researchers who do study them mainly do so from their own disciplinary perspective. Answering these questions is essential because the answers help us understand science better, reduce the constraints and biases we face when doing science and make new scientific breakthroughs. The answers also enable us to provide a comprehensive foundation for the field of science of science: a field that holds the potential to understand the foundations of science, its present boundaries and how to push those limits.

Improving science is very important for the simple reason that science shapes nearly all aspects of our lives. Scientific advances are behind all the technologies that dominate our lives, such as computers, smartphones, electricity and high-speed transportation that connect us to people and resources around the world. Science has provided us with theories that explain how our species evolved and the nature of matter and the universe. Science drives our unprecedented prosperity and has enabled us to reduce disease, malnutrition and poverty more than we ever have in human history. Yet science has also enabled us to endanger our own existence, through challenges like climate change and overpopulation. And we turn to science again to make new breakthroughs that can help tackle these challenges. During the coronavirus pandemic, scientists directly steered society and the decisions of governments, with much public debate about the role of science in our lives.

So it is key for our success as a species and for improving the conditions of our lives that we understand fundamental aspects of science: What is science? How did we start science? Why did science evolve the way it did? What forces drive science and new advances? What are the present limits of science

Science of Science. Alexander Krauss, Oxford University Press. © Alexander Krauss (2024).
DOI: 10.1093/9780198937401.003.0001

and how can we push those limits? It is key that we understand the what, how, when and why of science. But most scientists are generally busy doing science and do not have the time to study what drives science, the constraints facing the methods and instruments we use or how we can do science better. Scientists focus on producing results, evidence and theories. Rarely do they have the time to take a step back and study deeper questions about the foundations and present boundaries of science and what we know—about what enables and constrains the knowledge we acquire and the methods and instruments we use to acquire that knowledge. This book addresses these foundational questions about what makes science possible and how we can better conduct and advance science.

When attempting to answer these questions, most researchers take one individual disciplinary perspective—studying only the history of science,[1,2,3] scientometrics,[4,5,6,7,8,9] sociology of science,[10,11,12,13] philosophy of science[14,15,16,17,18,19] or psychology of science.[20,21,22,23,24] But science is a highly complex and multidimensional phenomenon and we cannot understand science from just one disciplinary perspective. To understand the environment and climate change for example, environmental sciences combine methods and evidence from ecology, physics, chemistry, physical geography, natural resource management, economics and atmospheric science. To understand the human body, medical sciences integrate methods and evidence from biomedicine, genetics, physiology, epidemiology, neuroscience, nutrition science and biostatistics. Taking an interdisciplinary approach is also the only way we can comprehensively understand a system as complex as science and its foundations and limits. There is no way around it. As we have not yet taken an integrated approach to studying science, we do not yet understand well what the foundations of science are and what drives science—at least not nearly as well as we understand the foundations and driving forces of the environment, the human body and other complex physical, biological and social phenomena. The best way we gain comprehensive knowledge is by using diverse methods from different fields that grounds our knowledge in those separate strands of evidence, as each method captures the phenomenon differently and strengthens our understanding when evidence is coherent across the different methods.

From an interdisciplinary perspective, when we look into any field studying science it can appear astonishing that the approach to studying science and the most studied and thus most important feature of science can vary so widely across isolated fields—and still be the 'most important' feature of science. For economists, the central feature has commonly been funding and incentive structures for rewarding and advancing science.[25,26,27,28] For statisticians, it has been the methodological constraints and biases in designing and improving our experimental studies in science.[29,30] For scientometricians and network scientists, it has been scientists' number of citations and publication records.[4,5] For sociologists of science, it has been the role of social influences on science.[10,11] For historians and philosophers of science, it has been the role of scientific theories including models,[1,14,15,18,31] and so on. These are all important aspects of science, but they have not yet been integrated together into a coherent account.

Consider the most well-known historian of science Thomas Kuhn and the most well-known philosopher of science Karl Popper. Kuhn dedicated his research to investigating the evolution of science, focusing on how scientific theories evolve. For him, science is not cumulative but undergoes paradigm shifts—that is, foundational changes in theories. Kuhn's account of science is arguably the most well-known account of science yet proposed. Despite widespread debate since Kuhn, the nature and growth of science remains elusive. Kuhn studied particular theories mainly in physics up to the early 20th century to develop his hypothesis of paradigm shifts.[1,32] Yet here we shift the focus across fields and to scientific methods and instruments used to do science, which illustrates the vast accumulation of scientific knowledge. Major scientific methods and tools used across fields, such as mathematics, microscopes and X-ray methods, and major scientific fields, such as physics, geography and genetics, have not been entirely discarded. And it is difficult for us to imagine that they could be. We instead constantly expand and refine them, which demonstrates the cumulative advancement of science.

Popper in turn adopted a different approach to the study of science and attacked Kuhn's account of non-cumulative science and constant paradigm changes, describing it as a 'lunatic fringe.' For Popper, the only scientific methodology needed and the defining criterion of what science is is whether we can falsify (refute) a scientific theory or not.[15, 33] This principle of falsifiability states: if we are not able to falsify a theory (such as string theory or a theory of astrology), then the theory is not scientific. This normative principle, for Popper, should define scientific investigation and the evaluation of theories. But it has not had much effect on the scientific community outside of philosophy of science. Falsifiability also tells us little about questions such as why science evolved the way it did and how we can best advance science in the future. Popper also argued that there is no logical structure of how scientific breakthroughs emerge.[34] Here we will provide integrated answers to such questions by adopting a holistic approach that goes beyond the perspective of any one discipline.

Consider also the work of network scientists who study science by leveraging the methods of scientometrics and big data. Scientometrics, a scientific discipline using large-scale data, studies science by analysing features of scientific publications and citation counts (how many times a publication is cited by other publications as a measure of its impact). Scientometricians including network scientists study topics such as scientists' collaborations, careers, productivity and networks.(4,5,8,9,35,36,37,38) This big data perspective to science has generated important insights about the dynamics and outputs of science. But it has not yet been able to explain well the general origins and limits of science and what drives science and the methods of science. These researchers are aware that the success of the field 'depends on us overcoming traditional disciplinary barriers' and are aware of the limitations of using scientometric methods to study how science works and advances: 'this bias toward citations is reflective of the current landscape of the field, [and] it highlights the need to go beyond citations as the only "currency" of science.'(4) cf. (35,37,39) Critics have also highlighted the dangers for the scientific community of overrelying on scientometrics that has generated a publish-or-perish system of adverse incentives and an increase in lower quality research across science.(37)

There is much research on science and these are just three examples among the range of different subfields studying science. These leading researchers, like those in other fields such as psychology of science and economics of science, provide important insights into our understanding of science. Yet each adopts their own perspective and focuses on their own area of research. This is observed looking at who they cite and who cites them—largely researchers within their particular subfield. No integrated account of how science evolves and advances has yet been developed that is coherent across fields. We cannot address most questions about the nature of science by adopting the common unidisciplinary approach—for example taking a scientometric approach that studies citation patterns. We depend on cognitive science to be able to explain how our cognitive abilities enable us to perceive, reason, do science and acquire knowledge about the world, and how our mind and senses present constraints on how we perceive and understand the world. We rely on methodology and statistics to be able to explain how we advance science by reducing such constraints and developing new methods and instruments, such as new experimental techniques and radio telescopes, that expand our scope to the world, and so on.

Because researchers studying science commonly address one aspect of science from one perspective,[37] this has led to blind spots in existing accounts of science. This book aims to shift this debate that, to date, has taken place within individual fields—not across fields. The interdisciplinary approach we take here enables us to understand the interconnected factors (methodological, social, cognitive, financial etc.) that make science possible and constrain science. As we will see, different factors influencing science, highlighted as highly important within a field, turn out to be much less important when compared against the range of factors across all fields. This has important implications on understanding what actually drives science most and on shifting the focus of our research in a number of fields.

Disciplinary specialisation is important and has provided much of what we know about the world. Yet we also need a meta-approach that pulls the disparate pieces together and provides us with an overall picture of science. *Only then can we establish which factors are most and least important across fields in understanding and advancing science, and identify the common mechanism underlying the different fields.* The central shortcoming in existing studies is that despite generally viewing the field from their own disciplinary viewpoint, researchers still attempt to explain, often entirely, what drives science. Here we broaden the scope of science of science by integrating the range of subfields outlined above and others that have been largely, or almost entirely, disconnected from the other fields studying science to date—from methodology of science,[30,40,41,42] economics of science[25,26,27,28,43] and computer science of science,[44,45,46] to cognitive science of science,[47,48,49,50] biology of science,[51,52,53,54] anthropology of science[55,56,57,58,59] and archaeology of science.[60,61,62) cf. (4,5,35] The central challenge of science of science is accounting for and integrating the existing empirical and theoretical knowledge and different perspectives from across disciplines into a holistic field. Only this way can we then build on that integrated knowledge. This has also traditionally been the central challenge towards greater understanding in related fields like philosophy of science. This book aims to address this challenge.

We will see here how our evolved mind and sensory abilities (to observe, experiment and process information) make doing science possible but also shape what and how we observe. Our scientific methods and instruments (such as statistics and electron microscopes) enable us to study a much broader set of phenomena, but also have constraints to how we measure them. Institutions, funding and societal challenges help influence what knowledge and research methods we produce, distribute and use. Scientific norms and methodological assumptions shape the way we evaluate our evidence, among other influences. A wide range of fields have insights that are very relevant in explaining aspects of science. This book is about fitting the pieces of this large puzzle together. Researchers studying science largely did not know that many pieces or even the puzzle existed. As we will see, some of the pieces of the puzzle are much bigger than others—that is, some fields are much better able to explain aspects of science than others.

This book aims to provide a unified framework for the field of science of science by combining methods and evidence from across the natural, behavioural and social sciences and thus offers a comprehensive understanding of how we drive knowledge and science, what shapes their limits and how we can improve them. The hope here is that the science of science will emerge as an integrated field that incorporates the range of subfields that study science, just as the environmental sciences and medical sciences are integrated fields that help overcome previously fragmented and disconnected areas of knowledge—which defines, to date, how we have studied and understood science. Within this integrated field, we can better comprehend the origins, foundations and limits of science. Understanding its limits is inseparably linked to understanding its origins and foundations, and vice versa. Identifying the most important features driving science and the limitations facing science is also important because it enables any scientist or scientific institution to direct greater attention to those features that are best at advancing science and to addressing those limitations. A better understanding of science can help scientists better identify bottlenecks and become aware of their own constraints and blind spots in doing science and how to address them. A number of insights provided throughout the book can even be directly applied by researchers in their own career—insights on increasing one's research productivity, questioning the assumptions one adopts and how one can develop new methodological techniques to do science better.

A central argument throughout the book is that to understand science comprehensively we need to integrate evidence of the abilities and conditions that have enabled us to develop science (biological, cognitive, social and methodological), the abilities and conditions shaping the scope of science (including, in addition, historical, economic etc.) and, most importantly, the abilities and conditions allowing us to expand the present limits of science (mainly methodological and instrumental but also cognitive, sensory and social). The account of science presented here explains how the scientific methods and instruments we develop play a central role in shaping the foundations and present limits of science by setting the boundaries of how we are able to observe, measure and experiment—that is, how we do science.

This methodological toolbox of ours sets the scope and present limits of what we can know and what is possible in science—while economic, social and historical influences help shape what we study within that scope and those limits. Methods and instruments we develop are the main mechanism through which we develop new knowledge and make new advances—while the role of other factors that can foster science (such as funding, incentive structures and the scientific community) vary widely; we observe in ground-breaking scientific publications that teams can be small or large, low or high funded, young or old, at low and top ranked universities, or interdisciplinary or not. On the one hand, we thus have our mind's internal methodological abilities to observe, experiment and process information (our *universal* methodological toolbox). On the other, we have the sophisticated external methods we have developed using our mind, such as statistics, radar telescopes and X-ray devices (our *adaptive* methodological toolbox). Taken together, our methodological toolbox has enabled us to develop and advance science, medicine and technology. We will show how scientific methods underlie the different factors explored across all subfields studying science. Our methods are the best way to bring together the different subfields and many are already interested in the important role of methods and tools. Given that the foundations and present limits of science are largely determined by our available scientific methods used to study the world, *a central focus on scientific methods is essential to the integration of the science of science*—as we will see.

The (methodological) tower of science: a holistic framework for science

We can think of science as a massive (methodological) tower of science that consists of three elements. The first element of the tower is its foundation at the bottom that is made up of human abilities—our senses and cognitive abilities—which enable us to observe the world around us and solve problems, test hypotheses, imagine and reason causally. As we are all born with these methodological abilities, we call this here our universal (built-in) methodological toolbox.

The second element of our tower of science are the different floors of the structure, and each major scientific method and instrument we have invented to improve our universal methodological toolbox represents a different floor. There is a floor on which the microscope is built, another on which statistics are built, another on which X-ray methods are built, and so on. The collection of all our methods and tools of science makes up the structure of the tower of science. This includes all scientific tools that vastly extend our visual scope of the world, such as microscopes, telescopes, X-ray methods and spectrometers. It consists of all quantitative methods that greatly enhance our cognitive abilities to process complex phenomena and study vast amounts of information, such as statistical and mathematical methods and controlled experimental methods. It encompasses all instruments that enable us to separate previously inaccessible substances like proteins, DNA and viruses, such as the centrifuge, electrophoresis and chromatography, and so on. Making such methodological innovations requires us to build on the foundation of the tower (our universal methodological toolbox) and on its existing floors—that is, on previous methodological innovations (our adaptive methodological toolbox).

This means that electron microscopes are useful as they enhance existing light microscopes, and light microscopes are useful as they vastly extend our evolved ability for vision. The same applies with our basic sense of quantity we have used to develop arithmetic, then statistics and eventually big-data methods, and so on. Each of these upgrades to our methodological toolbox immensely expands our ability and scope to study and understand phenomena—opening up the world of atoms, cells, molecules, ecosystems and galaxies. The first two elements of the tower make up the foundation and structure of science, without which we would not be able to conduct, develop and advance science. Without them, science would not exist.

The third element of the tower of science are the rooms of different scientific fields that are commonly made up and defined by the particular scientific methods and instruments that are applied. Each room is a body of knowledge we acquire using particular tools of science. That is, bodies of knowledge are (methodologically) organised in rooms. Cell biology for example was developed as a scientific field after inventing the electron microscope, and molecular biology after developing X-ray diffraction methods. A number of biology's subfields are on the floors of the tower that rest on microscopes, centrifuges and X-ray methods. Yet the rooms on a given floor also rest on other foundational floors below it, with methods like statistics and controlled experimentation near the bottom of the tower since most experimental research across science heavily relies on them. Complex knowledge and fields commonly build on multiple methods and instruments.

The shape and form of our tower of science has been continually extended over time. A few hundred years ago, the tower was much smaller. With the invention of foundational methods and instruments of the 17th century the central floors of the tower were built and enabled researchers to continue expanding the edifice of science. Each new major scientific method and instrument we design has amplified our scope and understanding of the world. Six new revolutionary methods and instruments developed around the 17th century brought about most major discoveries at the time: the first microscope was developed in 1590, telescope in 1608, barometer in 1643, air pump in 1659, statistics in 1663 and calculus in 1675 (Figure 0.1). The pioneering scholars Galileo, Hooke, Boyle and Newton and their contemporaries each applied one or more of these new tools of science to expand our understanding in astronomy, biology, physiology, pneumatics, mechanics and optics. These methods, including instruments, in their extended form, remain central and foundational to science today.

The tower houses the entire scientific community—about nine million researchers presently globally.[63] Funding also helps expand the tower and the shape it takes. It can foster the construction of new floors and rooms with new sophisticated methods and instruments and fill the shelves with new books of knowledge.

With each new major method and instrument we develop, a new floor is constantly being built onto the tower and existing floors are restructured. The shape and height of our tower of science is determined by the range of scientific methods and instruments we have invented thus far that enable us to study and understand the world. The strength of the tower is determined by the reliability of our best tools of science. We redraw the borders of the tower and thus the present borders of science as we expand our methodological toolbox, and they are also influenced by basic factors such as the size of the scientific community and available funding, especially in some research fields. Like filling the missing chemical elements of the periodic table of elements that make up the world, we continue to fill in more and more of the floors and rooms of the tower. But

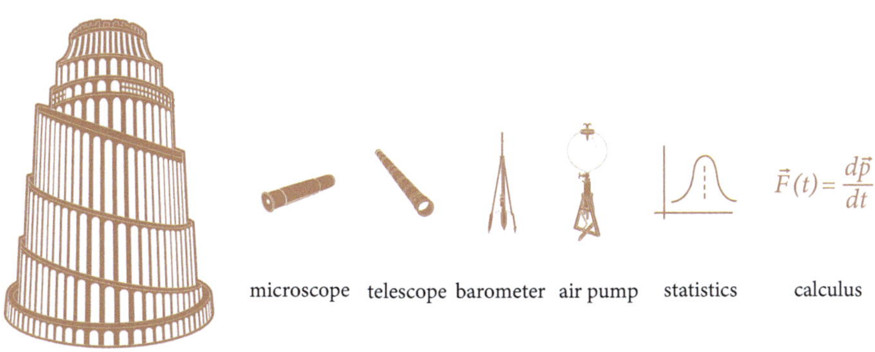

Figure 0.1 The (methodological) tower of science in the 17th century.

its future shape and form are not predetermined and we will later discuss the possibilities for expanding and stretching science in new directions.

The present shape of the tower of science is depicted here with central methods and instruments of science (Figure 0.2). These powerful tools have revolutionised science, making up the foundational and most important floors and pillars of the tower.

Our tower of science is today's version of Egypt's Great Library of Alexandria in the 2nd century BCE which aimed to gather all existing knowledge in the world. As we go through the chapters of the book, we will go on a tour of the tower of science, learning about how it is built, what conditions shape it, how it evolves and how we can improve it. As we will see, different subfields of the science of science provide different evidence and insights into the three elements of the methodological tower of science: the foundation of the tower (our universal methodological toolbox), the floors (our adaptive methodological toolbox) and the rooms (our scientific fields grounded in our methodological toolbox). To date, researchers who study science work on different floors and largely within isolated rooms—with most only publishing in journals within their own field. Yet studying science by making observations out of one window, as we will see, provides an understanding of science from one perspective. We will illustrate how to restructure the tower, so all researchers studying the

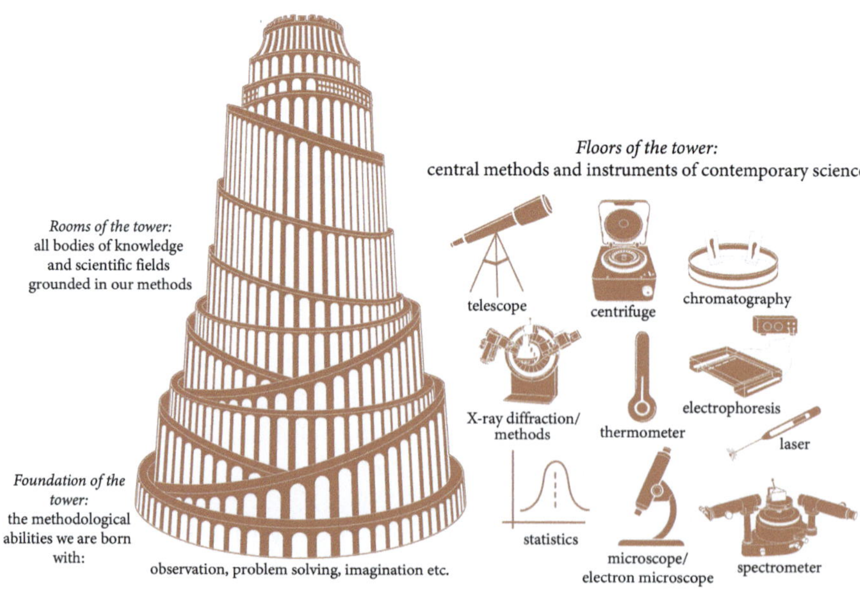

Figure 0.2 The (methodological) tower of science in contemporary science.

same topic—science—work on joint floors to gain a much more holistic view of science that complements and does not compete with each other (Chapters 16 and 17). Such an integrated approach will help us tackle our constraints and blind spots in the way we presently understand science. We will also later outline the different pathways for how we can further develop the tower in the future—that is, the pathways for the future of science and how we can push the present limits of science.

How the book is organised and the central argument

Let us provide an overview of the book's structure. We will first describe the integrated field of science of science by combining methods and evidence from 14 relevant fields that are grouped here into four areas: *internal factors* (biology, cognitive science, psychology and linguistics), *external factors* (economics and sociology), *historical and cross-cultural factors* (history, anthropology and archaeology) and *meta-level and methodological factors* (methodology, scientometrics/network science, computer science, statistics/mathematics and philosophy) (Figure 1.2 and Chapter 1). We identified these subfields of the science of science using the criterion of whether they help explain the origins, foundations or limits of science. Fields like physics and chemistry are not included as they are not as *directly* relevant; for science is a complex system in which we humans and the methods we apply are at the centre, and thus human sciences and methodological fields provide most insight, as we will see. By pulling these fields together, we provide here an overview of the literature across science that enables readers to gain a comprehensive understanding of science, on the one hand. On the other, it enables us to identify the foundational role of scientific methods at the centre of the field, explaining how methods are connected to factors across all 14 disciplinary perspectives (Chapters 2–15).

The ultimate test of a successful account of science is how useful it is in understanding science—in providing a comprehensive understanding of the foundations, boundaries and advancement of science. On this test, the holistic and methods-driven account presented here can take us further than individual fields of study alone (such as psychology of science, scientometrics and history of science) (Chapter 16). What emerges is an integrated and coherent theory, the *new-methods-drive-science* theory, that explains how we develop new scientific ideas and discoveries through the new methods and instruments we develop and apply collectively. The book aims to offer a unifying theory and foundation for the field of science of science (Chapter 17). With this integrated understanding of science, we then outline the present boundaries of science and how to push those boundaries. We also look forward and lay out the possible pathways that the future of science holds (Chapters 18–20). In the Conclusion we then draw implications for the field of science of science. We outline how we can help establish the field, how we can better measure science and discoveries with a broader set of methods and, most importantly, how we can foster ways to advance science.

In a nutshell, we can summarise the central argument of the book as follows: the different subfields that study science each capture one aspect of science, its evolution or its boundaries, but not the whole story. We need to integrate the different subfields to reveal the bigger picture of science. Each subfield has something specific to contribute to our understanding of science, and of scientific methods and instruments as the foundation of how we conduct and advance science (Chapters 1–15). To integrate the subfields, we need to better focus on scientific methods and instruments because they are the common thread where the subfields all overlap. When we then combine these different disciplinary perspectives, we learn that our methodological toolbox—which consists of our best microscopes, mathematical methods and X-ray methods—is the main driver of science by enabling us to perceive, measure and explain the world in new ways (Chapters 16 and 17). Within this methods-driven framework, we can see how the present limits of our methodological toolbox largely account for the present limits of science (Chapters 18 and 19). We need to thus expand our methodological toolbox to expand the present limits of science and the research frontier (Chapter 20).

In terms of the *research methodology*, explaining the foundations, limits and advancement of science requires adopting a cross-disciplinary framework that integrates methods from across the natural and social sciences. Evidence is derived here from studies using large-scale statistical analysis, experiments, surveys of scientists, historical analysis, big data analysis and other data sources. We take advantage here of studies using different quantitative and qualitative methods to examine the complexities of science. Each method provides evidence for different aspects of science. Importantly, the independent strands of evidence derived from the different methodologies are consistent with each other and, when pulled together, enable us to understand science more coherently and uncover which factors are most important.

While there are many advantages to quantitative studies, there are also cases in which a broader, synthesising study that adopts a qualitative approach is needed. To be able to provide a comprehensive overview of a multifaceted field like science of science requires pulling together insights from the diverse subfields studying science across the sciences and humanities. In doing so, qualitative studies enable exploring complex phenomena and patterns across those diverse disciplines. They can contribute to a holistic and nuanced understanding of the different factors influencing scientific processes and outcomes by integrating them together—from the evolution of ideas and methodologies to scientists' motivations and social norms. Qualitative studies can provide a bridge between disciplines and foster interdisciplinary understanding. Overall, they can enable us to better understand the broader context influencing the foundations of science. A companion book, *The Motor of Scientific Discovery*, analyses over 750 major scientific discoveries including all Nobel-Prize-winning discoveries and the methods and instruments used to make them and provides statistical evidence of how we drive science and discovery—and complements this book with such quantitative data.(64) Together, this qualitative and quantitative work enables developing this theoretical and conceptual framework that integrates diverse insights into the study of science.

Defining science and outlining the personal motivation for the book

A book on science needs to clarify what is meant by science, so we briefly do that here. The scope and complexity of science will become clearer as we progress through the chapters. Different views exist about what is involved in the seven-letter word *science*—with its Latin origin *scientia*, meaning knowledge. Commonly, science is defined as the study of the 'world through observation, experimentation, and the testing of theories,' according to the *Oxford English Dictionary*.(65) Yet scientists engage in many activities: they identify new research problems, design and conduct experiments, collect and analyse data, reveal new patterns in datasets, use imagination and analogies, develop explanations and theories, run simulations and make predictions, develop methods and instruments, generate models, visually represent phenomena, and derive implications for policy. Science is driven by many methodological approaches and redefined here as follows: science is the study of the natural and social world by using our cognitive abilities (including observation, experimentation and problem solving) and the methods and instruments we develop (including statistics and microscopes) with the aim of describing, explaining, predicting and controlling phenomena. Scientific *methods* include statistical techniques, controlled experimentation and algebra, and scientific *instruments* (tools) include particle accelerators, electron microscopes and electrophoresis (not other features of science such as concepts, theories and language).

A further clarification: since this book offers a different approach and views that are at odds with the disciplinary approach to studying science in existing studies, we briefly sketch out here the evolution of this book and its interdisciplinary and methods-driven approach. With my PhD completed, I worked with governments and the World Bank for five years doing research in applied statistics, economics and behavioural sciences. This included quantitative evaluations of the causal effects of public policies and reforms on topics like health, education and energy. Over time, I became more aware of the limitations, assumptions and biases of leading scientific methods and aware of the problems related to the causal results we derive from experiments and simulations. On the one hand, I increasingly began to question the foundations of the methods and evidence we use. I was constantly confronted with questions about what we can know with which level of certainty, what robust evidence is and how we can improve the way we do research. That is, I was confronted with questions about the foundations and present limits of our knowledge and the methods we use to develop that knowledge. On the other hand, I became aware that the common approach to scientific questions of using one method and disciplinary perspective will not take us very far in understanding and addressing complex challenges—whether malnutrition or climate change. These are commonly tackled from an interdisciplinary perspective in the applied research environment of governments and the World Bank.

I began to question why scientists in academia, in contrast, commonly approach a problem in a more narrow way rather than adopting multiple methods and an interdisciplinary approach that allows for more reliable and coherent evidence grounded in the different independent methodologies. It appeared that researchers' training in a given method together with constraints on time and resources drove the status quo. With the aim of addressing these questions about science I then transitioned back into academia. This applied research and concrete experiences have been central to adopting this book's big-picture approach that integrates methods and evidence across multiple fields.

Finally, this book is written with a broad audience in mind, including any academic and reader interested in understanding the origins, foundations and present limits of science. In education systems we are largely taught the outputs of science: facts, theories and laws about the world. We do not generally learn about the foundations of science, how science works (the process of science) and how to improve science. Yet with such an understanding, we are better equipped to broaden the scope of science and advance science more broadly. Within academia, this book could also be used as a textbook for a course on Science of Science, Metascience, Methodology of Science, Science and Technology Studies, History of Science and also Philosophy of Science. More than a general background and interest in science is not required and terms that readers may not be familiar with are defined as they are introduced.

1

Describing Science of Science

We first describe the existing landscape of the disparate fields studying science and then outline what an integrated and unified science of science looks like by providing a framework for the field. Science of science is an unexpected combination of terms at first glance. It involves scientists doing science to understand science. When studying science itself, we thus practice the activity being studied. We use scientific evidence to explain what science is and how it works. For methodologists and statisticians, it means studying the constraints and assumptions of scientific methods that they themselves adopt. For sociologists and psychologists, it involves studying the biases and norms in science that they themselves can be influenced by. For scientometricians and network scientists, it means studying publications and citation patterns in science by producing publications that they expect will be cited. For evolutionary biologists, it entails studying the evolution of our mind and its abilities to reason and perceive the world that they themselves have also inherited.

Given the paramount importance of science in society and about nine million scientists currently worldwide[(63)] it is surprising that there are only few dedicated full-time to studying science itself and how to improve science and its methods. It is surprising that science of science, as a discipline, did not develop as science developed. There are not yet interdisciplinary journals or university departments specifically dedicated to science of science. Yet across the various subfields that study science, the vast majority of existing publications cluster in five areas of knowledge, with the largest concentrations in philosophy of science (31%), history of science (25%), scientometrics/network science (16%), cognitive science of science (9%) and sociology of science (5%) (Figure 1.1). That these five subfields account for about 86% of all publications is a historical contingency. As we will see, they do not intrinsically provide a more comprehensive understanding of the foundations and limits of science than other subfields, like methodology of science and cognitive science of science. Each subfield contributes in part to a richer and more integrated understanding of science by studying a different aspect and providing one perspective to the larger picture. Overall, only about 3% of publications studying science use the term 'science of science' (or 'metascience'

Science of Science. Alexander Krauss, Oxford University Press. © Alexander Krauss (2024).
DOI: 10.1093/9780198937401.003.0002

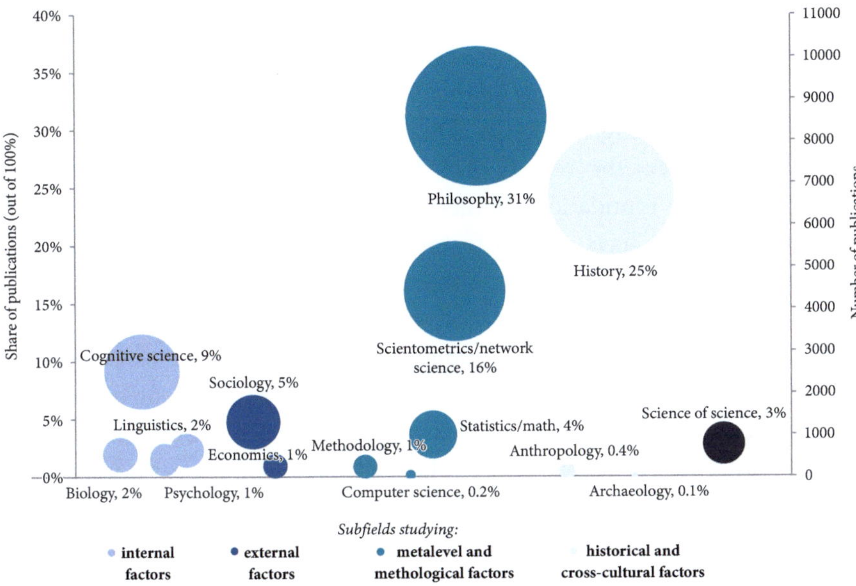

Figure 1.1 Share and number of publications across the subfields of science of science.

Note: data are derived from Scopus (the largest citation database of scientific journals), and reflect estimates for all existing publications—up to early 2024—within each subfield.(66) The estimated shares add up to 100% of publications across all 14 subfields including the 3% of publications using the term 'science of science,' 'metascience' or 'metaresearch' (on the far right). The shares provide a rough estimation of the distribution of research across fields provided in Scopus while they do not capture all publications in each research area. For interested readers, a note in the Appendix outlines how the shares are calculated.

or 'metaresearch') (Figure 1.1). As an indication of limited integration in the field, very few publications across all subfields use the common terms interdisciplinary, cross-disciplinary or multidisciplinary, at less than 3% in total. For example, within the over 8000 publications in 'history of science,' only 3.5% mention the term 'inter-/cross-/multidisciplinary' and 0.1% the term 'science of science.' Within the over 5000 publications in 'scientometrics,' only 12% and 0.6% mentioned these terms respectively.(66)

Different researchers studying science use a different method and unit of analysis and thus study different features of science. Leading scientometricians and network scientists like Fortunato, Wang and Barabási have focused on and stressed the key role of publications and citations;(4,5) leading historian of science Kuhn, the paradigm shifts in scientific theories;(1,32) leading philosopher of science Popper, the evaluation principle of falsification of scientific theories;(14,15,31) leading sociologists of science Latour, Woolgar and Bourdieu, the social practices of scientists;(10,11) and so on. Such disciplinary isolation has led to simplified and at times contradictory views. Leading researchers studying science have not studied how important, relatively speaking, the particular 'key' factor is that they study and how it relates to the other 'key' influencing factors in different fields which they do not study. We will later assess the role of each factor using the criteria of its scope in explaining the foundations, limits and advancement of science, and the direct influence we have on that factor in shaping science. Using these two criteria, we observe that these central factors proposed by the most cited researcher studying science within a particular field—namely Kuhn in history of science, Popper in philosophy of science, Latour and Woolgar in sociology of science, etc.—are not able to explain as much or have direct influence compared to other factors (Chapter 16). These leading researchers have overinterpreted the particular role of the factor they study compared to other factors, especially the foundational role of our scientific methods and our mind in enabling and constraining science that are not as commonly studied.

Classic work in the early origins of science of science goes back at least to Znaniecki in 1923,[(67)] Ossowska and Ossowski in 1935[(68)] and more generally to Galton's *English Men of Science: Their Nature and Nurture* in 1874[(69)] and later de Solla Price's *Little Science, Big Science* in 1963[(70)] and Zuckerman's *Scientific Elite: Nobel Laureates in the United States* in 1977.[(71)]

In Figure 1.1 we outline the landscape of existing research studying science and its concentration in particular subfields. In Figure 1.2 in turn we describe the unified approach to the field presented here that is needed to comprehensively understand science, combining the different bodies of research and methods which, to date, have been fragmented and isolated from each other. It provides an overview of what we will be covering in the chapters to come. The set of factors, and thus fields, are grouped here into four areas: *internal factors* (biology, cognitive science, psychology and linguistics), *external factors* (sociology and economy), *historical and cross-cultural factors* (history, anthropology and archaeology) and *meta-level and methodological factors* (methodology, scientometrics/network science, computer science, statistics/mathematics and philosophy).

Within the scientometric community that includes network scientists, the field has however been viewed narrowly as the 'field that relies on big data to unveil the reproducible patterns that govern individual scientific careers and the workings of science' by studying primarily publications and citations.[(4) cf. (5,35,39)] This common view among scientometricians illustrates

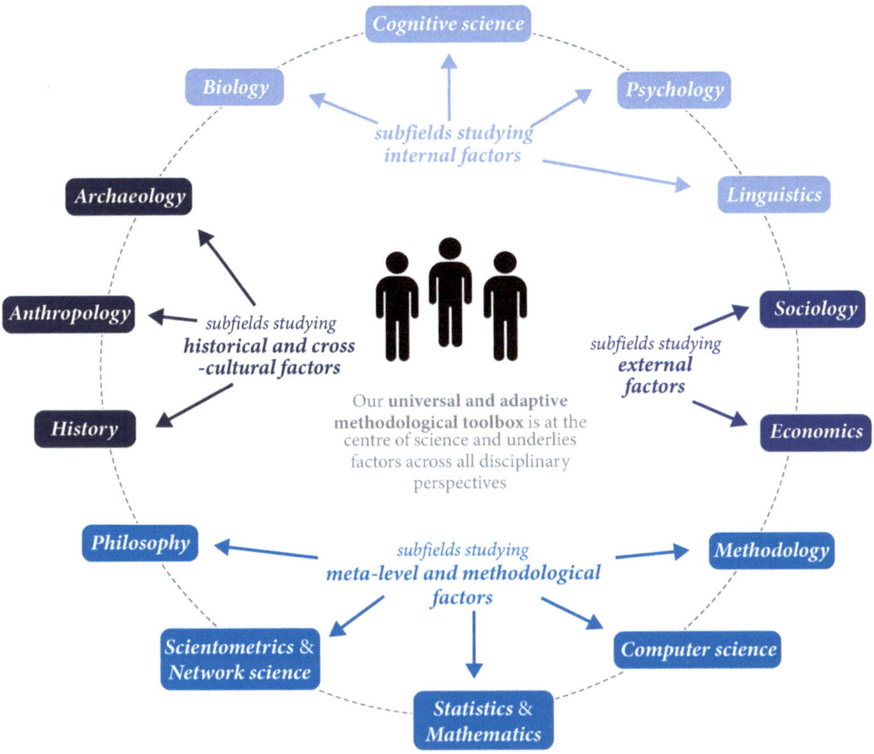

Figure 1.2 Science of science: explaining the foundations, limits and advancement of science from 14 subfields—an integrated field that combines multiple methods.

Note: this is the first study to demarcate the field of science of science by identifying the 14 relevant fields that contribute to our understanding of science and it groups them into four areas. Internal factors refer largely to the human body and mind; external factors refer to our broader environment; historical and cross-cultural factors refer to the past and different contexts; metalevel and methodological factors refer to metascientific aspects of science and scientific methodology. Other research domains that can provide insight into science of science can be categorised in one of these 14 subfields—for example public policy is included in economics of science, communication sciences in linguistics of science, data science in statistics, and the arts in terms of imagination and creativity in cognitive science of science. Finally, cognitive science of science broadly covers our evolved cognitive abilities and constraints (related to observation, memory and abstraction), while psychology of science narrowly covers psychological biases and personality traits. For interested readers, a note in the Appendix outlines which methods are commonly used in each subfield.

how they view the field in terms of one method. Though they recognise that the success of the field 'depends on us overcoming traditional disciplinary barriers,'[(4)] there are a range of other disciplinary methods, evidence and perspectives from across the other fields studying science that are not taken into account. Scientometrics has however dominated the research studying science in leading multidisciplinary science journals to date.[4,5,35,37] Figure 1.2 outlines what an integrated science of science, without disciplinary divisions, looks like. The integrated field presented here can be defined as follows:

> The field of science of science is the study of science, and especially the foundations, limits and advancement of science and scientific methods, that integrates methods and evidence from across the natural, behavioural and social sciences.

The field addresses foundational, methodological and meta-scientific aspects of science. It studies fundamental questions that span the scope of the field: What drives science? How do we develop science? What constrains science? How can we improve and advance science? What are the structures, processes and dynamics that underlie the production and evolution of scientific knowledge? More specifically, what particular factors drive scientific progress and innovation? Many factors like methodological and technological advancements, collaborations, diversity of perspectives, funding and incentives are still not fully understood. How can we more accurately measure and evaluate the impact of scientific research? Current metrics, such as citation counts, have many limitations and do not capture the immediate impact of new ideas in science or the broader impact of research on policy or society. How can we better ensure that science has a positive impact on policy, society and people's lives? How can we improve and incentivise the replicability of scientific research? How do cognitive biases and social factors influence scientific research, such as the interpretation of data and the publication of results? How can we minimise these biases and promote more objective and rigorous research? What are the ethical implications of scientific research, especially in fields like gene editing and artificial intelligence? How can we better integrate interdisciplinary approaches in scientific research? How can we promote more effective communication and dissemination of scientific research, including on social media and digital platforms? These are some of the important questions in the science of science.

The field assesses the methods and instruments of science, the process of science, how we design, implement and evaluate scientific studies, and domain-specific topics from its subfields. This is done by applying a range of methodologies. These include empirical studies, experiments, surveys of scientists, historical analysis, big data analysis and conceptual analysis, which are integrated across domains and subfields. The field studies science from across disciplinary borders—and from both high altitude and the bottom up. The objective of the field is straightforward: by better understanding the foundations and present limits of science and scientific methods, we can do science better and drive new knowledge and discoveries. This is the field of science of science. And given that scientific methods are at the foundation of science—developed and used to conduct, advance and improve science—*we need to place scientific methods at the centre of focus to integrate the science of science.* No factor other than our scientific methods and instruments is implicit in all 14 factors and fields and can we more directly influence and improve to do science and make new breakthroughs.

Interdisciplinary fields like environmental sciences are excellent examples of why we can only understand a complex, multifaceted subject by adopting a range of methods and evidence from across fields—just as in the case of science of science. Environmental sciences are the study of environmental problems. And defining, analysing and addressing such problems are only possible by taking natural, social, economic, political, historical, psychological and other dimensions into account simultaneously.[(72)] The natural dimension of environmental sciences involves studying environmental problems that arise in our natural environment through physical, chemical and biological interactions and that are linked to our awareness of those problems and how we measure them. Similarly, science of science studies how we measure nature using our methods and instruments and how we have evolved in our natural environment that has shaped our abilities for observing, solving problems, causal reasoning and acquiring knowledge. The social dimension of environmental sciences studies how population growth and cultural norms can lead to environmental problems. Similarly, science of science studies how population size, cooperation, specialisation and societal influences shape science and how we conduct science.

The economic dimension evaluates the role of natural resources, their depletion and their interrelationship with economic growth. Similarly, science of science evaluates the incentive structure of scientists and how economies of scale foster cumulative knowledge. The policy dimension assesses how we can mitigate environmental challenges and how international institutions can coordinate initiatives. Similarly, science of science assesses the public institutions that help plan, finance and manage how knowledge is produced, distributed and used. The historical dimension analyses the environment on a large timescale that includes geological and human history. Similarly, science of science analyses the evolution of science, discoveries, methods and their complexities throughout history. The psychological dimension examines how citizens perceive environmental challenges such as climate change and the level of willingness to change their behaviour. Similarly, science of science examines our psychological biases and the scientific methods developed to reduce them. Assessing the evolution of global warming and simulating its future course requires using sophisticated statistical programmes and computer technology. Similarly, science of science uses complex statistical and computational methods to assess large datasets of discoveries, scientists and publications. Importantly, environmental sciences, like science of science, requires applying multiple methods to be able to capture the multiple factors at different levels and thus multiple dimensions of the field.

The emergence of the environmental sciences enabled a much more comprehensive understanding of the environment by developing integrative methods like the dynamic integrated climate and economy (DICE) model. It has enabled us to evaluate climate change and its multiple environmental, social, economic and historical dimensions more realistically than with previous isolated assessment models. For developing this integrative approach William Nordhaus received the Nobel Prize. No such integrated methods and analyses yet exist in science of science. Yet, commonly, attempting to understand the multifaceted phenomenon of science from one perspective is like attempting to understand the environment by just studying oceans or changes in temperature. We learn a lot, but the approach is deficient in providing a comprehensive understanding.

Just as this interdisciplinary approach has already been adopted in such fields, it is just as feasible and necessary to adopt it in the science of science. We outline here a vision of science of science in which many, if not most, researchers studying science can continue to pursue disciplinary specialisation, but all would begin to spend a share of their time to be broadly informed about existing research on the same topic across the other subfields of science of science. This way we can ensure our understanding of science is more coherent—that is, coherent across the subfields of science of science. Just as environmental sciences arose as an integrated field in the 1960s and 1970s to be able to understand complex environmental problems,[(72)] the aim of this book is that science of science arises as an integrated field driven by the same need for a cross-disciplinary approach to be able to understand the complex nature of science and discovery. In the next chapter we begin by assessing the evidence derived from biology as we work through the different topics that help us understand aspects of the foundations and limits of science.

2

Biology of Science

What are the evolutionary origins of science and how can they help us understand how we do science today? Like other animals, our ancestors evolved abilities for vision and other senses that enable us to perceive the world, and evolved other related physiological functions. These provide insights into science's evolutionary origins. Species like ours require making observations and acquiring knowledge about the world to be able to survive and meet basic needs. This requires knowing about what foods they can and cannot eat, and about their ecological environment and other animals. They have to identify regularities in nature.(60,73) Chimpanzees for instance forage using different types of tools and techniques. They crack nuts using stone and wood hammers and extend their reach and extract termites, ants and honey using sticks (Figure 2.1). They gather water using leaf sponges, throw stones as weapons and use levers for different tasks.(21,74,75,76) Using tools to solve problems, chimpanzees thus have a toolkit that they acquire through social learning and experimenting. To use such tools requires chimpanzees to have a clear objective of the tool in mind, predict how the tool can enable them to achieve that objective and understand how the tool must be applied. They thus must comprehend the interactions needed between the physical tool, their hands and the desired outcome, which illustrates their ability to acquire knowledge and manipulate the world around them. Tool use is widespread among many animals, from crows to sea otters and octopuses, that manipulate objects for their purposes.(74,77)

Chimpanzees are also able to think of reality abstractly and create and test hypotheses about the behaviour of others, when they for instance deceive others.(21) Non-human primates also reason about objects, space, quantities and the mental states of others, use classification systems and recognise causal relationships.(21,74,78) Omnivorous mammals—from rats to humans—for example, reason causally by perceiving similarities and patterns in the world and basing their behaviour on them. They make for instance the connection between food with a peculiar taste and gastrointestinal sickness occurring afterwards.(79)

Science of Science. Alexander Krauss, Oxford University Press. © Alexander Krauss (2024).
DOI: 10.1093/9780198937401.003.0003

Figure 2.1 Chimpanzees reason causally when using tools like sticks to 'fish' for termites.
Source: Mike R., Wikimedia Commons.

We can also better understand our constraints to studying the world by comparing them with the evolved abilities of non-human animals. In some ways, these abilities surpass our human abilities—and, in other ways, they have abilities that we humans do not.[(21,52)] Different birds for example use magnetic fields to orient themselves, perceive their altitude and create mental geographic maps.[(74)] Certain chimpanzees have extraordinary working memory, particularly flash memory. This allows them to quickly memorise things within less than a second (far exceeding humans), as experiments with chimpanzees and humans illustrate.[(21,74)] Ultraviolet-seeing insects such as bees are directed towards the centre of many flowers that have developed ultraviolet streaks, allowing bees to find nectar and foster pollination.[(80)] Echolocating bats perceive their surroundings, including moving prey, using sound waves.[(50,74)] We humans did not evolve these and other abilities to the same extent. The sensory abilities of any species (including ours) thus only pick up parts and fragments of information and energy in the world compared to all the sensory abilities across millions of species.[(81,82)] We have a limited scope to the world. To support our limited perception, we have developed methods and tools. These enhance different aspects of how we can perceive the world—as we discuss in Chapter 10.

The parts of our surroundings that we, with our evolutionary adaptations, are able to observe and sense do not thus reflect the only way to view reality. Different species, with their own set of adaptations, survive using their particular means of perception. Bees, gorillas, condors and humans each distinguish between types of plants and animals in different ways, and classify geographic areas and their species' behaviour in different ways.[(81)] There are different ways to perceive and categorise phenomena in the world given that there are different species in the world.

In fact, while doing science or even just reading this book, our brain cells require sufficient water and food to function properly and retain information, and we require sufficient sleep and warmth to be able to concentrate. We depend on our environment and physical resources to be able to survive and acquire knowledge about the world around us.

There is another biological dimension of science: our human perspective to the world—as the biological animals we are—shapes what knowledge and objectives we pursue as we use our mind and the methods we develop to do science. Just by being human and members of our species, we direct more attention to some phenomena than to others, namely phenomena that fall within the environmental niche of the world we have evolved in and live in. Nearly all scientists study aspects of reality relevant to our needs and wants—human biology, human technology, human society, human diseases, human behaviour and other problems and objectives we humans face. Science is human-driven.

Large science funding agencies (like the European Commission, National Science Foundation and other public funding bodies) generally require researchers to outline the human impact of their research to be able to receive funding. Most science funding worldwide is spent on studying human beings, with for example 52% of total public research funding in the US allocated to medicine/health, life sciences and psychology. The remaining 48% is allocated to all other disciplines that also generally aim to benefit human beings and human progress, including engineering, physical and social sciences, environmental science and computer science.[(83)] So why do tens of thousands of scientists worldwide work on explaining and predicting illnesses, pandemics, population dynamics, financial markets, our behaviour and weather but only few scientists on explaining and predicting phenomena such as dark energy of the universe, the flora and fauna of the earth's deep oceans and the mind of insects? For one more directly impacts our lives while the other does not.

And scientists study the latter topics often to the extent that it may benefit and help better understand ourselves. We humans are our own point of reference. We want to enhance our conditions and lifespan but not generally those of all living organisms, including all animals and plants, unless they benefit us—as evidenced by the vast majority of research across the biological, medical, behavioural and social sciences. Overall, we study the world from our human perspective, our anthropocentric context, that shapes the present scope of science.

In sum, smart animals are able to meet their needs and create knowledge using their biological abilities for vision and their other senses that lay the basis for our methodological abilities that we use to be able to do science (Chapters 3 and 4). Understanding these common abilities that our species shares to different degrees with other species illustrates how we are able to develop knowledge and thus what abilities have enabled us to start science.[82] These abilities make up the foundation at the bottom of the *tower of science* that we will be reconstructing as we go through the chapters. These abilities thus provide insight into the foundations of science and help us understand aspects of our current scientific landscape, including our research focus. But biology of science *alone* is (just like scientometrics, psychology of science or any science of science subfield) incomplete. This is because science is shaped by our mind and the methods and instruments we use to do science, while also influenced by our broader social, economic and historical context.

3

Archaeology of Science

Archaeological artefacts that include increasingly sophisticated tools developed by early humans provide evidence for the origin of science. They offer historical evidence of the evolution of our methodological abilities to reason and acquire knowledge that are needed to construct those artefacts and to do science today. We humans have evolved abilities to observe, solve problems, experiment, categorise, reason causally and test ideas or hypotheses.[(22,74)] These, together, account for our methodological abilities of the mind we use to be able to develop knowledge and make sense of the world around us. Using these abilities, early humans such as *Homo erectus* and Neanderthals created complex tools such as hand axes at least about 1.5–2 million years ago.[(84,85)] Making tools like hand axes requires the ability to imagine and plan what they will look like before creating them. Early stone toolmakers needed to make mental representations, inferences and predictions. Making stone tools requires the ability to systematically observe, experiment, reason causally, test a hypothesis, imagine and plan in order to produce a clear preconceived object.[cf. (86,61)] Early humans demonstrated an incredible advance in controlling fire about 600,000 to 1 million years ago and developing sophisticated fire-hardened spears about 400,000 years ago.[(60,87)] This requires reasoning complexly, interconnected inferences and evaluating hypotheses.[(88,89)] These are also the evolved human abilities commonly used in contemporary scientific practice.

To understand the foundations of science, we have to understand the evolution and abilities of early hominines. This can present a challenge in drawing conclusions from fossils and material artefacts about our early species' abilities.[cf. (90)] Because the past is beyond the scope of experimentation and because our ancestors' cognitive abilities and methods do not fossilise, our understanding of the past and their specific abilities can be constrained. Yet we can extend our reach into the past and partly explain it by combining historical evidence we do have with current evidence.[(91)] Given that we know early human species developed such complex tools[(60)] and given that for us to develop them requires a systematic approach and refining the tools, we know they also used these methodological abilities (Chapter 4). No other way exists

Science of Science. Alexander Krauss, Oxford University Press. © Alexander Krauss (2024).
DOI: 10.1093/9780198937401.003.0004

to explain the foundations of how early humans reasoned and gained knowledge other than our methodological abilities of the mind they all required using.

Turning to our species, we *Homo sapiens* are the evolutionary product of millions of years and slowly emerged as a distinct species in Africa an estimated 250,000–300,000 years ago.[21,90] The way we perceive and view the world is shaped by our evolution. The ability to abstract, imagine and use analogies developed in our early ancestors over time.[49] This essential ability is used by contemporary scientists when reasoning, using methods to represent information about the world and developing theories. Cave paintings, sculptures and other symbolic art created to represent phenomena abstractly during the Palaeolithic can be seen in some ways analogously to models created to represent phenomena abstractly in science.[60] Niels Bohr for instance used our solar system as an analogy to develop a quantum model of the hydrogen atom, imagining electrons in the hydrogen atom behaving like planets orbiting around the nucleus. Charles Darwin modelled and illustrated his theory of evolution using the branches of a tree, with all species being related and humans just on one branch of the same tree.[51]

Abstraction and modelling reality are fundamental to how we reason and simplify the complexity of the world around us. We use representational models of the world in most scientific fields—from theories and analogies, to experiments, statistical simulations and mathematical equations. Our cognitive and social abilities for greater cooperation, language, abstraction and imagination have developed in a symbiotic way. They have built on each other through continual cumulative feedbacks that have made these abilities increasingly complex over time,[52,21] together with our abilities for developing more complex methods and tools (Chapters 4 and 10).

Combining our abilities, we have become better able to develop a remarkable set of increasingly complex methods and tools. We created for example bone harpoons and a range of flake tools to hunt and cut things more easily. We built dwellings to provide protection from the elements of nature.[60] We created what are viewed as early systems of notation, by using an ordered set of spatially distributed marks engraved on stones and bones to record, process and pass along information.[62] We produced symbolic drawings and depictions to help document observations and provide an external source of memory.[60] Developing such methods and tools requires (as in the case of constructing a dwelling) the cumulative use of systematic observation (to identify the best wood, mud and stones for construction). It requires causal reasoning (to understand that a well-constructed dwelling will protect us from rain and predators) and experimentation (to identify better construction methods). It requires imagination (to conceive how the dwelling will look before constructing it) and collectively planning within groups (to collect the materials needed for its construction). These are the human abilities we use across contemporary science, though more explicitly. Our early methods and tools can be seen analogously to those we develop today—such as microscopes and telescopes to extend our visual abilities (Chapter 10).

We eventually, using our methodological abilities, learned how to domesticate animals, which requires us to have knowledge of biological reproduction, the nutritional needs of animals and selective breeding techniques to foster particular traits. We learned how to cultivate crops, which requires us to understand the causal interactions between seeds, rain, soil fertility and erosion and often knowledge of annual cycles. It requires us to continually experiment with seeds to produce more productive yields and involves collective knowledge about selective planting, storing seeds and often about irrigation methods.[92] Early farmers had comprehensive botanical, zoological and ecological knowledge and were able to control parts of their environment through agriculture and livestock—as hunter-gatherer groups commonly still do today.[21,56,87,88]

As societies became more complex, we increasingly specialised in different professions and crafts. An estimated 9000 years ago in the Middle East, architectural improvements led to the development of rectangular buildings constructed out of stone and mud bricks that were often two storeys. Some of these well-designed buildings exist today. Such construction requires us to understand geometric dimensions, the weight that different materials can support and the need to heat limestone to temperatures of 750–850 °C to be able to produce plaster floors.[93] Constructing them is not possible without extensive collective planning, systematic measurement and experimentation. Such specialised professions enabled us to develop new bodies of knowledge that cumulatively built on existing knowledge. In the same way, without large communities of specialised professionals and cumulative knowledge, contemporary science would not exist today.

With the emergence of the first civilisations, we made large leaps towards science.[82] We developed the earliest known systems of written language and mathematics around 6000 years ago.[92] The shift from oral to written systems marked an important transformation that enabled us to use our cognitive and methodological abilities increasingly systematically and explicitly. These systems make recording what we observe easier. They reduce our cognitive constraints in processing and remembering information, making mathematical calculations and building on existing knowledge.[21]

The development of systems of writing reflects a pivotal historical shift, not only in the cumulative development of science but also in understanding science. Written language has enabled us to better comprehend the scientific and technological advances of our ancestors, as they were able to describe them in written form. Without a written record, we are limited to only studying other archaeological artefacts.

In the civilisations of Mesopotamia, Egypt, China, Central America and India, astronomy was for example a common area of knowledge developed using systematic observation and written records. It involved sophisticated astronomical models for making reliable predictions.[94] We learned that we can use the sky as a clock for time keeping, a compass for orientating ourselves and a calendar for planning agriculture, weather and temperature. Despite acquiring this understanding, our early ancestors could not yet explain stars, comets, the sun and other astronomical phenomena that largely remained intellectual puzzles.

Early written use of geometry can be traced back to around 5000 years ago in Mesopotamia and Egypt. It involved principles of areas, lengths, angles and volumes that were identified empirically and used for surveying, agriculture, construction and astronomy.[95] Ancient Egyptian civilisation and Norte Chico civilisation (modern Peru) constructed vast pyramids at least 4500 years ago (Figure 3.1). Building a pyramid requires—both then and today—applying principles in engineering, architecture and geometry that are grounded in systematic measurement, planning and experimentation. Farming in such civilisations was improved by experimenting using controls, namely testing different crops and comparing the outcomes, and thus acquiring knowledge by experimenting comparatively. An Egyptian medical

Figure 3.1 Pyramids of the Norte Chico civilisation constructed using systematic measurement, planning and experimentation, circa 2600 BCE.
Source: Kyle Thayer via Wikimedia Commons.

textbook from about 1600 BCE provides detailed cumulative experimental knowledge of dealing with injuries, fractures, tumours and various surgeries and it applies the methods of examining, diagnosing, treating and prognosis.(96)

A controlled experimentation is for instance described in the Old Testament.(97) The book of Daniel (1: 12–13) outlines an experimental trial with control groups that tests the influence of a vegetarian diet: 'Test your servants for ten days. Give us nothing but vegetables to eat and water to drink. Then compare our appearance with that of the young men who eat the royal food [and drink wine], and treat your servants in accordance with what you see.' We require combining our cognitive abilities to test such a hypothesis and conceive the design for such a controlled experimentation. Causal reasoning is required to test whether a potential cause (in this case a vegetarian, water-based diet as opposed to a non-vegetarian, alcohol-based diet) has an observable effect on people's physical appearance. Carrying out a trial is required, including systematically recording and comparing the outcomes of the physical appearance between the two groups after 10 days, and then deriving inferences from the trial outcomes to modify people's diets in the future. We then combined controlled experimentation with further methodological features such as randomisation and blinding in the 19th century (so that participants do not know which group in an experiment they are assigned to and are randomly allocated to groups) which further reduces human bias.(98)

Overall, developing such technological, agricultural and architectural innovations requires us to manipulate the world by applying our methodological abilities of the mind that we have used collectively throughout human history, including in contemporary science. Developing such innovations throughout history—and even just surviving—requires us to be able to understand and predict phenomena in the world. This involves knowledge of the characteristics of physical objects and forces (intuitive physics), the characteristics of animals and plants (intuitive biology), the behaviours and views of others and our own (intuitive psychology),(99) the characteristics of places, directions and shapes (intuitive geometry)(100) and distinctions between quantities (intuitive arithmetic).

In sum, contemporary science builds on these intuitive conceptions of the world and on our mind's methodological abilities that are evidenced in vast archaeological records. These have enabled us to reason and develop knowledge and tools, and they provide evidence of what abilities have enabled us to start science. The foundation at the bottom of the *tower of science* is grounded on these methodological abilities, which we will see in the next chapter also shape the scope of how we do science today. Archaeology of science accounts for one piece of the foundation of science and thus of science of science. Understanding these methodological origins of science, as we will later see, is important to understanding the current dynamics of science and how we can best advance science.

4
Cognitive Science of Science

Our mind makes doing science and creating knowledge about the world possible. It allows for vision needed to make observations, memory to recall what we observe, language to express what we observe, and reason to solve problems and develop scientific methods. We have evolved these methodological abilities of the mind over time and we use them to be able to study, experiment and acquire knowledge of the world. Our mind thus shapes how we get by and make sense of our natural and social environment, on the one hand. On the other, we face cognitive and sensory constraints imposed by nature and evolution, and also constraints on the methods and instruments we develop using our mind. They set the scope within which we are able to do science and create knowledge about the world. Cognition thereby refers to our ability to perceive, process, retain and act on information from the world around us.(101) Because our mind makes reasoning and creating knowledge possible, we can best understand our mind's abilities by studying how they have evolved to enable us to reason and create knowledge the way we do.

Our cognition, like all features of human anatomy and other natural phenomena, is the outcome of our evolution. It is the outcome of a largely unpredictable path of trial and error over the past few million years. This basic fact helps understand our cognitive abilities and their limitations: our mind has evolved largely reacting to problems we have faced up to now, within our environmental and cultural niche.(21,48,52,87,102,103) Our mind has developed over much of our history within the savannahs and natural landscapes around us. It has evolved by making sense of those parts of reality (especially people and our surroundings including plants and animals) that are relevant for our species' needs and survival and that we can access with our senses.(81,104)

Our mind has thus evolved over our species' history in large part for the pragmatic purpose to help us meet our basic needs. Our evolutionary history is one that did not directly involve observing and mentally modelling phenomena such as the size and nature of the universe, the historical origin of life, global financial markets and the emergence of conscious experience. Only in more recent human history and only by creating methods and instruments have we been able to develop such complex theories about phenomena

Science of Science. Alexander Krauss, Oxford University Press. © Alexander Krauss (2024).
DOI: 10.1093/9780198937401.003.0005

that we do not have direct sensory experience of and that generally do not directly affect our biological fitness. This has been one of the great mysteries of our mind and science:[105] how have we, given our evolutionary history, evolved the ability to do science and develop highly elaborate knowledge about the world? We have illustrated how we have evolved cognitive abilities to systematically observe, solve problems and experiment that enabled our species to develop stone tools, shelters and eventually agriculture and helped us meet our basic needs. Creating sophisticated scientific knowledge today is made possible by using these same evolved abilities but for purposes that do not directly influence our survival, as when they first developed. These abilities also include abstraction, imagination and creativity which we have developed over the last few hundred thousand years within social groups (Figure 4.1).[21,60,78]

We experience the world through our senses that send signals to our brain which creates an image of the world around us.[50,103] Through our senses, we perceive the world directly (in an unaided way), as three-dimensional and generally as well ordered by expecting future phenomena to resemble past phenomena.[21] We collectively develop methods such as statistics to study phenomena in the world using observational data that requires us to create variables (using values such as 0 or 1) to be able to calculate statistical relationships for those observations—which are often viewed as one-dimensional (Chapter 13).[42] Our mind and the methods we create using our mind determine what we are able to observe in the world and the way we are able to process what we observe.

Many phenomena in the world—given our cognitive and perceptual limitations—fall below or lay beyond the directly observable conditions in which our mind and senses have developed. The further we move away from these conditions—from the surface of the earth, from our ecological niche,

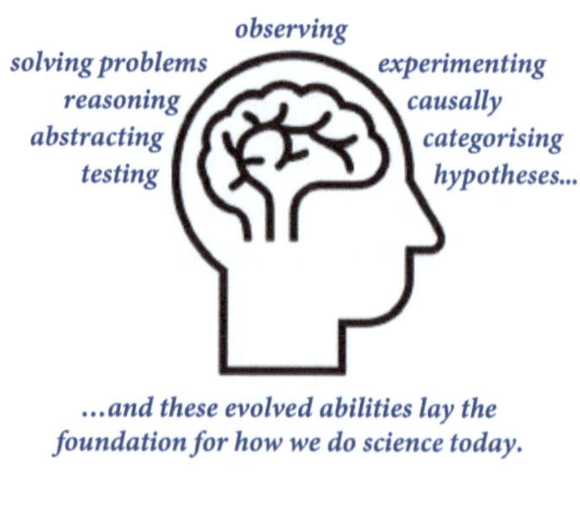

Figure 4.1 Our mind has evolved a range of abilities that we have used more systematically over time.

from our particular context—we generally require greater abstraction of the phenomena we study. Our mind is generally not able, without the aid of methods and tools, to access most phenomena studied in science beyond our senses—when we move to minuscule phenomena such as atoms and photons at the quantum level, vast phenomena such as magnetic fields and gravitation, extremely fast phenomena such as the speed of light, extremely distant phenomena such as the earth's core and planetary systems, or unobservable social phenomena such as global economic markets and political systems that lie beyond the observable actions of its actors. The same applies to many phenomena accessible to our senses, such as biological functions and chemical reactions that we have observed throughout our history. But to understand them systematically we have had to first develop new methods and instruments to access them. Just as our evolved visual abilities do not allow us to observe very small or large phenomena in the world, our evolved cognitive abilities do not allow us to process large sets of observations or understand highly complex phenomena well. Methods and instruments we have created collectively using our flexible mind explain most of the expansion of science by enabling us to study phenomena that would otherwise lie beyond our basic cognitive and sensory reach (Chapter 10).

We create knowledge either using only our methodological abilities of the mind (to observe, solve problems and experiment) or applying scientific methods and instruments that we can only develop using these abilities. Our methodological abilities thus *always* drive the knowledge we develop—while we do not always apply the tools of science that we create using these abilities. Our tools of science are products of our methodological abilities that we use to better study the world or solve problems more effectively. Our methodological abilities, scientific methods and knowledge also interact through a cumulative bootstrapping process in which they extend each other, in increasingly complex ways.

What is observable and non-observable also importantly shapes the knowledge we develop. Our vision, by allowing us to observe phenomena around us, provides our main and most used sensory input to make sense of the world (including for other primates).[50,106] Vision is a fundamental biological adaptation that we—and most animals—benefit from. It enables us to extract information about the position, movement and features of objects around us. It is through a narrow range of radiation in the form of light that we thus acquire most knowledge about reality.[50,106] Our eyesight, together with instruments we create to enhance our vision, set the scope and bounds within which we are able to observe phenomena. Our visual experience is the most common form of evidence we use to explain things. Human experience and understanding of reality is thus strongly shaped by the objects that fall into our visual range.[107] Theorising and visual perception are also generally inseparable in science. To theorise about something, we generally fall back on observations of the phenomenon we are studying or observations of related phenomena and then infer or extrapolate from them. Even theories developed in fields like theoretical physics and theoretical economics thus rely on observations of the world.

We cannot directly observe with our own eyes or sense most phenomena we study—from physical forces, fields and subatomic particles, to proteins, neural signals and complex economic systems. It is the methods and instruments we create that enable us to access and study them. To improve our ability to observe more easily accessible phenomena, we have created microscopes and telescopes (Chapter 10). Even for phenomena not directly observable we generally develop visual representations to better understand them, ranging from models of DNA and diagrams of chemical compounds, to figures of statistical distributions and economic graphs.

We generally understand phenomena more abstractly as we move from what is observable to what is non-observable. In the behavioural sciences, we can see this in the difference between studying the observable actions of people from a behavioural perspective and studying their unobservable mental states from a neurological perspective which are more difficult to understand. In physics, we can see this in the difference between Newton's focus on the more easily observable aspects of the world and the less observable world of quantum mechanics which is more difficult to grasp. When we study phenomena not observable with our sensory organs and not directly testable, we require greater mental abstraction, imagination and interpretation.

Our perception enables us to conceptualise not only observable phenomena but also non-observable phenomena. Light for example is viewed as acting as both a particle and a wave, depending on the measurement methods and experiments we apply.[81,108] And as we humans have not evolved to be able to perceive or conceive phenomena with both these properties at the same time, our cognitive ability to conceptualise phenomena at the quantum level is constrained. The peculiarities of wave-particle duality are the product of a complex interaction between how the world is and how we are capable of perceiving and thinking about it.[81,108] It is because we can imagine phenomena in the world as being a particle or a wave that we can create theories about non-observable phenomena at the quantum level possibly possessing both of these properties. We thus describe non-observable phenomena based on what we know about observable phenomena or what is conceivable using information about observable phenomena. We are not generally able to conceptualise non-observable phenomena in ways that are fundamentally different from our observation-based concepts (such as particles or waves). And when we do conceptualise phenomena differently—for example abstractly as statistical variables for medical treatments, mental states or inflation—we generally alter the dynamic phenomena to some degree. This is because we capture them in quantitative variables that are amenable to statistical analysis. To conceptualise phenomena like dark matter and global economic systems, because they are not directly accessible with our retinas or senses, requires greater abstraction than studying phenomena like rocks, plants and people.

Because many topics that scientists study—from theoretical physicists to theoretical economists—are not visible and we cannot sense them, the challenge of making *sense* of parts of the world is often related to our limited human senses and the ways we can observe phenomena using methods and instruments we develop to enhance our senses. This helps explain the greater limitations that confront theories in theoretical fields given the constraints we often face in collecting data to test and verify them—such as string theory and theories of multiple universes (Chapters 18 and 19).

In sum, we are only able to do science and create knowledge about the world by using our methodological abilities of the mind for observing, solving problems, experimenting and abstracting. Our mind, sensory abilities and evolutionary niche shape the knowledge we develop today. Cognitive science of science lays a fundamental basis of science of science. The central foundation of the *tower of science* rests on these methodological abilities that shape the scope of science and the scientific methods and tools we create to extend these abilities and thus perceive and measure the world in new ways—the topic of Chapter 10.

5
Psychology of Science

When we do science and acquire knowledge about the world, our mind also faces psychological biases. We face limited mental resources, time constraints and incomplete information, so we use simplified heuristics like rules of thumb or shortcuts when reasoning.[102,109] We for example often rely on existing assumptions and evidence when formulating a hypothesis or applying a given scientific method rather than questioning and testing them every time. In general, our mind has largely evolved to be able to absorb and process a limited amount of information and then make quick assumptions, decisions and conclusions based on that (incomplete) information. We think fast, are habit-based and use heuristics most of the time. This can result in unconscious biases.[109,110]

Experiments illustrate that scientists often have limited knowledge of their own unconscious biases in their work. A study of ecology scientists showed that they have low awareness and understanding of the importance of biases and how to mitigate them.[111] A study of forensic specialists illustrated that they generally viewed their own judgements as almost infallible and they demonstrated limited understanding of cognitive biases.[112] A study of medical doctors illustrated that they regularly made errors in clinical practice due to cognitive biases, and such biases were made throughout the diagnostic process when collecting, processing and confirming information.[113]

In general, our human cognitive constraints can present biases at all steps of the scientific process. This includes when designing and conducting experiments, such as confirmation bias when searching for evidence consistent with the hypothesis being tested. It includes when analysing data, such as omission of some results and poor understanding of statistical methods used. It also includes when writing up results for publication, such as HARKing bias (hypothesising after results are known) and confirmation bias when only reviewing existing literature consistent with the tested hypothesis (Chapter 13).[114] We are thus more likely to accept new evidence if it supports our already held views and theories (confirmation bias).[114,115] Researchers at times develop a new hypothesis or modify an existing hypothesis after analysing the results of a study—that is, they present an ex-post hypothesis as

Science of Science. Alexander Krauss, Oxford University Press. © Alexander Krauss (2024).
DOI: 10.1093/9780198937401.003.0006

an ex-ante hypothesis (HARKing bias).[116] Experiments show that we often accept existing evidence and fall back on the same decisions and assumptions we made in the past (status quo bias). Our expectations can help shape the outcome of our observations by influencing interpretations we make and information we use (expectancy bias).[50,109] Our expectations and hypotheses can at times shape our results by making us more inclined towards observations and outcomes that fit them and more easily rejecting those that do not.[117]

Such biases can appear to be evolutionary design flaws—given their empirical inaccuracy in different contexts. Yet we can better think of them as cognitive features adapted to quickly solve problems under resource and time constraints that have often been important despite trade-offs for less accuracy.[102] The way we reason and do science, when viewed as evolutionarily adaptive, is not always unbiased and accurate. In science, 'we all find it difficult to see the flaws in our own work—it's a normal part of human cognition.'[115] Yet we can reduce some of our individual psychological biases when reasoning by using methods, such as statistical techniques, randomisation and blinding, designed to mitigate such biases (Chapters 10 and 13). We can also reduce such biases through greater awareness, peer review, pre-registration of study designs and independent replication of studies by other researchers (Chapters 6 and 13).

Our reasoning is influenced not only by psychological biases but also by personality traits.[118] Drive and discipline foster systematic reasoning, just as curiosity, creativity and analogical reasoning foster how we do science.[119] As the historian of science Thomas Kuhn stated: 'A man may be attracted to science for all sorts of reasons. Among them are the desire to be useful, the excitement of exploring new territory, the hope of finding order, and the drive to test established knowledge ... Many of the greatest scientific minds have devoted all of their professional attention to demanding puzzles,' that is, to 'solving a puzzle that no one before has solved.'[1] And Einstein famously said: 'I have no special talent, I am only passionately curious.' Intellectual stimulation and recognition can also provide motivation to solve a problem or develop a new theory. Goals and needs influence and motivate us. And personal interest and social contribution can coincide for mutual gain.[118,120]

Newton's seminal book *Philosophiae Naturalis Principia Mathematica* in 1687 had the greatest impact on physics for several centuries.[121] And Newton made continual efforts to receive priority for his work, including in his manuscripts at least 12 different defences for his precedence in developing calculus before Leibniz whom he charged with plagiarism.[120] James Watson, who helped develop the theory of the structure of DNA, portrays himself as aggressive and arrogant in his book *The Double Helix*.[122]

He depicts competition as a defining trait of his thinking and research, and is largely motivated by the desire for fame.[(120)] Watson even later decided to auction off his Nobel Prize medallion for $4.1 million in 2014.[(123)] In his autobiography, the Nobel laureate in physics Max Planck also expressed his sense of competition and wrote that he had 'the desire to win, somehow, a reputation in the field of science.'[(124)] Darwin wrote that 'It seems hard on me that I should lose my priority of many years' standing' in reference to Alfred Russel Wallace developing a similar evolutionary theory at the time.[(120)] Since scientists build on the work of others and attempt to go beyond that work, scientific advancement is often grounded in a symbiotic relationship between cooperation and competition. Competition can help ensure quality control and independent testing of others' work. While a desire for recognition is a basic human trait, the sociologist Robert Merton highlights that most renowned academics are known to be driven by a desire for fame—from Galileo and Descartes, to Faraday and Freud.[(120)]

Scientists are not a special breed of Homo sapiens but are driven by goals, interests and personality traits like everyone else. A belief that scientists may be more clever or quicker at solving problems (or more objective or ethical) than people pursuing other professions is largely that, a belief—as a comparative study found among engineers, neurosurgeons and the general population that put into question the common expressions that 'it is not rocket science' and 'it is not brain surgery.'[(125)] This is connected to the widely held 10,000-hour hypothesis in psychology that states that about 3.5 years of full-time training (8 hours times 365 days) are needed to excel in most human domains including a scientific field.[(126)] Overall, scientists are not at birth entirely different from other people (who do not wear white coats or do not have the label PhD after their names).

In sum, we face psychological biases that influence our reasoning and our scientific results. Though, we can reduce them through research methods designed for that purpose. Personality traits also influence our motivation for doing research. Reasoning and acquiring knowledge are however not mind-bound but occur in our social and physical world. To better understand them, we need to study not only our individual biological, cognitive and psychological constraints and biases within us—that is, internal factors that largely revolve around our methodological abilities of the mind (Chapters 2–5)—but also the collective methods we develop using these abilities and the range of external factors that include social, economic and historical influences (Chapters 6–15; Figure 1.2). We have to thus study both the foundation of the *tower of science* and the factors that influence the shape and form of the tower.

6
Sociology of Science

If the cognitive abilities of children at birth a few hundred or even thousand years ago and today are likely not very different, what can explain the large differences in their theories of the world as adults? While our mind's methodological abilities to observe, solve problems and experiment are a precondition of science and we have used them more systematically over time, changes in broader demographic, social(21) and economic factors have fostered developing vast knowledge and methods. There were hundreds of scientists a few centuries ago.(127) Today, there are about nine million full-time scientists worldwide.(63) Demographic growth and complex social organisation have been crucial for the scientific community to grow and for greater collaboration, cumulative knowledge and methodological development (Chapters 10 and 11). Doing science is thus not just a cognitive activity conducted by individuals (Chapters 2–4). It has become an increasingly complex social activity conducted among a community of researchers.(21,78,87,88)

After birth we are socialised into a system of language and mathematics and we acquire much of our knowledge of the world through institutions designed for that purpose—schools and universities. We are raised in a cognitive and social environment full of information. We do not just use cognition but culturally embedded cognition when we reason and do science.(128) Without formal education and methodological training we would not be able to develop highly complex knowledge (Chapters 10, 12 and 13).

We are not disinterested and detached observers of reality but guided by institutions, shaped by scientific norms and motivated by values that can influence our research. The scientific community we are embedded in influences which problems, questions and objectives we find relevant to pursue and how we frame them. It shapes which methods we consider credible for analysing them. It influences which experimental designs we choose, the way results are assessed and what assumptions are allowed (Chapter 14). It shapes how we classify phenomena and how we define variables for them that influence how we measure and understand phenomena. It influences how we use results in medicine, technology and society. Our scientific community also sets norms

Science of Science. Alexander Krauss, Oxford University Press. © Alexander Krauss (2024).
DOI: 10.1093/9780198937401.003.0007

such as the kinds of evidence accepted (and not accepted) to support a claim, and the types of hypotheses and theories that are suitable (and not suitable). It sets norms such as the kinds of information that needs to be included (and not included) for publishing articles, and the forms of peer review that are appropriate (and inappropriate) for publication in journals. Such norms are ingrained in our scientific communities and help direct our scientific activities. They are defined differently across fields and change over time (Chapter 8). Together, they account for the rules of the game for doing science—rules that each player needs to abide by if they are to stay within their field. Acquiring knowledge through systematic observation, robust methodological analysis, significant results and reliable conclusions thus first requires agreeing on common criteria for evaluating what counts as systematic, robust, significant and reliable.

Researchers across and within fields also adopt different methodological approaches to studying the same phenomena. US-American scientists for example have adopted a more narrowly focused approach and understanding to genetics and natural selection than German scientists, who in contrast view evolution as also operating at the macro level.(129) In science of science, the study of topics like collaborations among scientists or scientific impact is approached using different methodologies. Scientometricians including network scientists commonly adopt a descriptive empirical approach, economists a causal empirical approach, sociologists largely a qualitative empirical approach, philosophers a conceptual approach, and so forth. This often leads to different answers to the same question as approaches are not yet integrated.

The sociologist of science Harriet Zuckerman published the seminal book *Scientific Elite: Nobel Laureates in the United States* in 1977. Zuckerman embarked on a journey travelling across the US to interview Nobel laureates about their personal background, family and research.(71) Her pioneering research investigated the demographic and social traits of eminent scientists, including attributes like age, sex, religion and ethnicity. She describes the process of how being 'a Nobel laureate is, for better or for worse, to be firmly placed in the scientific elite.'(71) A central argument of hers is that science is a collective effort that is embedded in a social and cultural context, with the scientific elite made up of talented individuals influenced by social and institutional factors like education, social stratification and mobility.cf. (119) She emphasises the role of power, authority and influence related to science's award system. Studying these ultra-elite of science, Zuckerman suggests that the process of discovery is highly competitive and hierarchical, in which only a privileged few can ascend to a high status. Her research also underscores that Nobel laureates are dedicated to their work, often at the detriment of other facets of their lives, with many laureates working tirelessly, at times over decades, in search of a single breakthrough over the course of their careers.[71]

Robert Merton—a central inspiration and later collaborator of Zuckerman[71]—was another eminent sociologist of science. In his influential book *Science and Technology in a Democratic Order* in 1942, he argues that science and scientific advances take place within a scientific community with shared scientific norms, values and institutions.[130] Merton identified the influential 'Matthew Effect' in science, which states that renowned scientists receive much more credit for their research than less renowned scientists with equally important research contributions.[131] The social structure of science thus generates a self-perpetuating cycle of scientific success among established scientists. He also highlights the broader relationship between science and society, exploring how scientific advances can have important implications for technology, social change and policy. In 1993, at the age of 83, Merton married Zuckerman, with the duo leaving behind a large mark in the sociology of science.

Two other leading sociologists of science, Latour and Woolgar, investigated the role of social influences within the laboratory.[10] They observed that scientists within a leading biological laboratory are exposed to peer and social pressures and seek influence. And influence is not just achieved by the theories they develop but by the scope of their social networks and their ability to mobilise support for their work.[10,11] In attempting to explain science, they argued that social influences are the most important factor in creating knowledge—though they only focus on one factor in studying science, namely social influences. In the book *Science of Science and Reflexivity*, another leading sociologist of science, Bourdieu, argued that science may be 'in danger of becoming a handmaiden to biotechnology, medicine, genetic engineering, and military research,' given the possible risks and interests of corporations.[11] Science can in turn also bring negative effects for society and the environment, by contributing to global challenges like overpopulation and climate change. Bourdieu's aim—similar to Latour and Woolgar—is 'to identify the social conditions in which science develops.'[11] But by focusing almost exclusively on social conditions, they strongly overvalue the role of social influences in science—while neglecting the role of methods and our mind, and the broad range of other interrelated factors (biological, historical, economic) that help drive science (Chapters 2–15).

In studying science, the highly influential sociologist of science Latour had very limited knowledge of science—in fact, 'Latour's knowledge of science was non-existent'[10] and he thus adopted a largely naive, anti-science position that lacked knowledge of most other features of science. Since then, much related work in the sociology of science has taken such a constructivist stance that has been parodied[132] and rejected[133] as it is not supported by rigorous empirical evidence in other fields of science. Such work has done more harm to the perception of studies of science than contributing to our understanding of science. And it is a case in point for why we must take the range of cognitive, methodological, demographic and other factors into account when analysing science to avoid misattributing the role of a single factor—social influences like power and peer pressure. This strong position also further constrained disciplinary integration that can be traced to the so-called science wars that further divided sociology of science from history and philosophy of science, and embodied vast disagreement about what the study of science should entail.

The scientific process is not only influenced by social factors but is itself a social process in which we generate knowledge by building on each other's work and methods. We read the articles of others. We apply methods and instruments developed by others. We discuss and verify our results with colleagues, peer reviewers and editors. In science, we observe organised curiosity (in co-authored publications), coordinated critique (in peer review) and public funding for research teams (through institutional grants).

No single person can develop a complex method, field or theory on their own. A complex method (from statistics to randomised controlled experimentation) has not been created by an individual mind but by many individuals building on the work of others. An academic field (from molecular biology to nuclear physics) has always been developed by collectively working together in a cumulative way. A complex theory (from quantum theory to a theory of the origin of life) has not been created by the mind of a single person without relying on much previous knowledge. To create such knowledge we need to acquire and share information cumulatively—we need cumulative knowledge. And the larger a scientific community, the more researchers can cumulatively build on each other's research.

One of the greatest discoveries of the 20th century was uncovering DNA's double-helix structure in 1953. Commonly, the names of two scientists are mentioned when referring to the discovery: Francis Crick and James Watson. Yet the discovery built on initial work by Miescher, based on the pivotal X-ray work produced by Franklin and Gosling, without which producing the image of the double helix would not have been possible. That in turn required applying X-ray diffraction methods developed by von Laue, who used X-radiation identified by Röntgen. The work was also supported by parallel research on DNA structure by Wilkins and his group of colleagues, among many others.[122] Rosalind Franklin arguably contributed the key missing piece by applying the method of X-ray diffraction using DNA fibre to be able to identify the structure of DNA (Figure 6.1). Crick, Watson and Wilkins however won the Nobel Prize (directly after Franklin's death) for the work that builds on her research. The discovery of the molecular structure of DNA revealed how information is transferred in living material and opened vast new areas of research.

Another social dimension of science are the gender disparities we observe across scientific fields, with science remaining heavily biased towards males. Assessing all over 500 Nobel-Prize-winning discoveries shows that only 3% of all Nobel laureates in science are women.[64] Physics is particularly male-dominated, with only 2% of Nobel-Prize discoveries in the field made by women. The share is 6% in astronomy and 7% in medicine. While shares remain very low, we can observe a positive trend, as more than half of all female Nobelists who were ever awarded the Prize received it since 2000.[64] A number of major discoveries have however been in large part made by women who did not receive credit or a Nobel Prize for their work. Rosalind Franklin is a classic example. Another example is the microbiologist Esther Lederberg. She made the discovery of a virus which infects bacteria, and with her husband

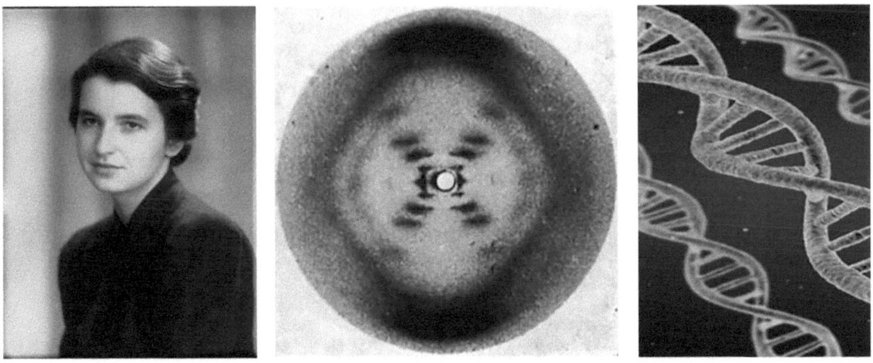

Figure 6.1 Discovering DNA's double-helix structure was the outcome of multiple scientists working together, with Rosalind Franklin (left) and her X-ray 'photo 51' (middle) playing the key role in revealing the double-helix structure (right).

Source: Elliot & Fry/National Portrait Gallery (left); Raymond Gosling/King's College London via Wikipedia (middle); Qimono via Den Store Danske (right).

created a technique to transfer bacteria between petri dishes. But in 1958 her husband Joshua Lederberg received the Nobel Prize for the research carried out with his wife, and he only mentioned her once in his Nobel lecture.[(123)] Another example is the astrophysicist Jocelyn Bell Burnell who discovered pulsars in 1967, and she published an article together with Antony Hewish. But in 1974 Hewish was awarded the Nobel Prize for the work.[(134)]

An explanation for the very low levels of female Nobel laureates is that women have been systematically discriminated in accessing education and science throughout history. The unfavourable norms about the role of females in science have begun to improve since the second half of the 20th century and especially in the 21st century. But women still remain especially underrepresented in STEM (science, technology, engineering and mathematics) fields. We also still often hear sexist statements in science. The Nobel laureate and biochemist Sir Tim Hunt for example said in 2015: 'Let me tell you about my trouble with girls. Three things happen when they are in the lab: you fall in love with them, they fall in love with you, and when you criticize them they cry.' Hunt was heavily criticised by social media for such sexist statements.[(123)]

Moreover, knowledge across science is not just created or discovered, but must also be explained, justified and generally replicable. When for example Darwin had an initial idea about natural selection or Einstein about a possible relationship between mass and energy, they did not yet necessarily produce reliable knowledge. Knowledge gains reliability and validity when others independently arrive at the same results using the same and other methods, others can replicate the results and we can justify the methods used to others (Chapters 10 and 14).[cf. (135)] Theories require gaining acceptance and consensus among members of a scientific community. And that is not always straightforward for some major new discoveries when they are first made. Quantum theory was at first a particularly controversial discovery in physics, which motivated Max Planck to polemically claim that: 'A new scientific truth does not triumph by convincing its opponents and making them see the light, but rather because its opponents eventually die, and a new generation grows up that is familiar with it.'[(124)]

In sum, we develop complex knowledge, methods and science within a scientific community and social context. Science cannot be entirely independent of our scientific practices, norms and institutions. These help coordinate how we produce and distribute new knowledge. Newton in 1675 said that 'If I have seen further, it is by standing on the shoulders of giants.'[cf. (136)] And his claim holds for contemporary science. Returning to the tower of science analogy we see that sociologists of science are among the most isolated from other researchers studying science on different floors of the tower and they take their own perspective and publish research in journals within their own field. Later we will illustrate how to restructure the tower so all researchers studying the same topic, science, work on the same floor to gain a much more holistic view of science that complements and does not compete with each other. Social and demographic features play a crucial role in understanding and fostering science, and thus in science of science. And they are closely connected to economic features.

7
Economics of Science

We can foster science through economies of scale, a reward system, science policy and targeted research funding.[25,26,27,28,43,137] As society becomes more productive, diversified and efficient in providing goods and services, more individuals can dedicate themselves to scientific activities. Larger communities of scientists have a comparative advantage over individual isolated scientists in cumulatively building on research. Science can function like an economy: just as a growing and more specialised labour force generally develops more diversified goods and technologies, a growing and more specialised scientific community generally develops more diversified knowledge and methods. Economies of scale and agglomeration facilitate greater division of labour across and within scientific fields—and thus greater methodological diversity and knowledge.

Science runs on a priority-based reward system that motivates research and innovation.[71] It works by giving priority to the first person to publish a new idea or develop a new method. It also requires us to make research and methods publicly available.[25,28,138] As a form of intellectual property right, priority is rewarded through social recognition from the scientific community and through the potential of contributing to society (Chapter 5). This winner-takes-all system thus incentivises scientists to produce and share knowledge,[71,120] and generally more so than monetary incentives.[25]

Public institutions also help plan, finance and manage how we produce, distribute and use knowledge. They intervene in how we produce different areas of knowledge by setting government priorities, science policy and resource allocation—with funding also influenced by overall economic development. They shape how we distribute knowledge through schools, universities and research institutions and how efficient that knowledge is mediated.[28,139]

To practise science we generally require some level of funding. During the centuries between Newton and Einstein, research was commonly conducted on low budgets and a small scale. As Einstein put it, 'Science is a wonderful thing if one does not have to earn one's living at it.' Scientists funded themselves or received small funds from universities or private foundations. Then companies like Bell Telephone and General Electric began to conduct basic

Science of Science. Alexander Krauss, Oxford University Press. © Alexander Krauss (2024).
DOI: 10.1093/9780198937401.003.0008

research in their laboratories. It took governments until around the time of World War II to realise that science had driven the enormous advances in technology, medicine and human well-being up to that point. This is when public funding for science began to be strategically targeted for the first time in a coordinated way.(140) The Chinese Academy of Sciences was grounded in 1949, the National Science Foundation in the US in 1950 and the German Research Foundation in 1951. In the US for example, total annual expenditure on science was negligible before the war, accounting for 0.1 billion US$ in 1930. After the war, the US became the most important player in science funding and output. There was a nine-fold increase in total annual expenditure in research and development in the US, increasing from 1.5 billion US$ to 13.6 billion between 1945 and 1965.(141)

Since most science spending today is public and funded by taxpayers, we can view much scientific research as a public good that needs to be made publicly and freely available.(25,26) Scientists do not generally have at their disposal large amounts of funds to conduct research in certain areas of science, like basic research, cost-intensive research or research only yielding returns after many years.(5,25) The costs of laboratories, equipment and running certain experiments in some fields within chemistry, biology, medicine and especially physics and astronomy are at times high for individual researchers to bear on their own.(5,25) Without government funding, some areas of research would not be possible or would be underinvested. Developing some of the most expensive scientific instruments in the world would not be feasible, such as CERN's (Conseil Européen pour la Recherche Nucléaire) large hadron collider—the world's largest particle accelerator—which cost billions of dollars (Figure 7.1). The collider led to the discovery of a new particle, the Higgs boson, in 2012 which is a central part of the standard model of physics.

There are however large shares of discoveries that have been made with only limited funding. Assessing over 750 major discoveries, including all Nobel-Prize-winning discoveries, we observe that hundreds of discoveries have been made using low-cost methods and instruments, such as statistical and mathematical methods, light microscopes, electrophoresis, assay techniques, thermometers, chromatography methods and centrifuges. These are among the ten most commonly used methods and instruments in science and today we can acquire them new for less than a thousand or even a few hundred dollars.(64) Large-scale funding is today not always needed to make novel breakthroughs. The most expensive scientific instruments are in contrast concentrated especially in subfields of physics and astronomy. They include the massive Hubble space telescope that enabled discovering the accelerating expansion of the

Figure 7.1 Large-scale public funding is needed for some expensive scientific instruments, such as CERN's large hadron collider (particle accelerator).
Source: CERN.

universe in 1998. The massive laser interferometer gravitational-wave observatory (LIGO) led to the discovery of the existence of gravitational waves in 2015. The Manhattan Project required developing new isotope separation techniques and reactors which led to the invention of the first atomic bomb in 1945. Exceptionally expensive projects in genetics and medicine have also invested vast amounts of funding targeted to cancer research and the human genome project. These well-known examples of the most costly scientific instruments and initiatives required vast government funding. Such large-scale funding is fortunately not the norm but the exception, and hundreds of discoveries are made using low-cost methods and instruments.

Funding agencies have also increasingly required research to have social impact—for example to influence outputs like diagnostic tools, technologies, medications and public policies.[27] This can shape the supply of research by determining which scientific areas receive more funding. For some areas of research, funding can shift in response to societal challenges that we face, such as climate change, energy efficiency and health pandemics. Funding can also shift as scientific challenges emerge, such as the replication crisis in science and big data. Research agendas are often set to meet the needs and objectives of those funding the research—mostly wealthy societies. For example, diseases more common in wealthier countries such as diabetes and cancer receive much more research funding than tropical diseases that are potentially much easier to cure but mainly affect those in poorer countries with less financial resources to fund science. Private funding in science, through pharmaceutical companies for instance, is also biased towards the demand of wealthier countries. Governments and funding agencies have become, in most fields, increasingly focused on science that produces quicker results through an incentive system based on shorter-term outputs, especially articles. This can lead to a shorter research horizon that neglects long-term research projects (Chapter 11).[25,27] How resources in biomedicine within the US are allocated for example is more strongly associated with existing allocations and research than with the present burden of diseases, illustrating a mismatch between allocated resources and actual needs.[142]

Importantly, some studies, especially by economists, have begun assessing the causal effects of interventions on questions in science of science. A study carrying out a field experiment evaluates how search costs affect new scientific collaborations by randomly assigning 402 scientists to different sessions (sharing information on grant funding opportunities) at a research symposium. It finds that scientists assigned to the same session (the treatment) were more likely to build a new collaboration for a grant at 28% than scientists assigned to different sessions (the control) at 16%.[143] A study using a regression discontinuity research design evaluates the Matthew effect ('the rich get richer') in science funding among 3660 grant applicants. It finds that researchers do not win subsequent grants because of greater achievements but rather because they already won a previous grant as a younger researcher, which suggests that receiving funding early on provides a strong advantage for receiving later funding.[144]

A study using a difference-in-differences research design assesses whether the premature death of 452 prominent scientists changes the evolution (measured by publication rates and funding flows) of subfields that they published in before passing away, compared to control subfields. It finds that when eminent scientists pass away, fields more easily evolve in new directions.[145] A study conducting a randomised controlled trial (RCT) assesses the effects of open peer review by assigning reviewers to either a group who signed their reviewer report (revealing their identity to authors) or a group who did not sign the report (remaining anonymous) for 408 manuscripts. It finds that signed reviews had higher quality, required longer time to complete and were more likely to recommend publication than unsigned reviews.[146] A greater shift towards conducting such studies with research designs that estimate the causal effects of interventions driving science is a critical step to move the field of science of science forward and bring greater rigour—which we will discuss in the Conclusion.

In sum, economic factors and public institutions shape funding priorities, research objectives and science policy. A reward system also helps provide some incentives for scientists to develop new scientific advances, discoveries and the methods needed to do so. Economic factors influence the research and scientific tools that we can fund and thus the size and shape of the tower of science. But economics of science *alone* is—like any science of science subfield—incomplete and cannot explain the foundations, limits and advancement of science. For science is largely shaped by our methods and mind, while also influenced by our broader social and historical context.

8
History of Science

Science has a history, and when we study the history of science including the discoveries, theories and the methods used to create them, we can trace their origin to the individuals who developed them. Thomas Kuhn, the most cited and well-known historian of science in the 20th century, offered an explanation of the history of science that rejected the view of scientific change as being cumulative.[1,32] The history of science can be viewed as a cycle in which established ideas and facts are doubted, new problems and evidence then lead to new revolutionary ideas and facts (and replace the established ones), which eventually over time are also doubted once problems and anomalies associated with them become apparent, and the cycle begins again. For Kuhn, this process of science is not cumulative. It represents revolutionary paradigm shifts, in which a scientific community rejects existing assumptions, concepts and theories and adopts entirely new ones. Kuhn thus describes science as going through paradigm shifts. For him, 'revolutionary' scientists help redefine which research questions and assumptions are important and what methods are best. In short, they help rewrite the rules of the game. This notion of science may seem to apply to shifts in physics in the past, namely shifts in our theories of physical reality from Aristotle to Newton and then to Einstein. Kuhn focused on such cases largely in physics up to the early 20th century.[1]

The shift from the Ptolemaic earth-centred theory of the universe to the Copernican sun-centred theory characterised the classic paradigm change, which Kuhn focused much research on. In geology for example the shift from a stabilist to a plate-tectonic mobilist view of the earth can also, according to Kuhn's thesis, be viewed as a paradigm shift. In economics, it can reflect the shift from a neoclassical view of individuals acting as perfectly rational actors towards more behavioural approaches that view individuals as acting irrationally at times. In physics, it can reflect the shift from a classical Euclidean view of time and space as flat to an Einsteinian view reflecting the curvature of time and space. And it can reflect the shift from a classical mechanics' view of physical reality in which particles act in a deterministic way (Newton) to a quantum mechanics' view of physical reality in which phenomena act as both a wave and a particle at the same time and in a probabilistic way (Bohr,

Science of Science. Alexander Krauss, Oxford University Press. © Alexander Krauss (2024).
DOI: 10.1093/9780198937401.003.0009

Heisenberg and others). Each of these changes in theory has been made possible by changes in the methods we use to study the phenomenon. Such changes in theory in the past can appear as large shifts in science. But they have become very rare and are in fact based on much cumulative knowledge, as we will see. Quantum theory, the mechanisms of evolution, computer science, microscopes and the like have all been continually and extensively refined over time.

No major scientific methods used across fields (such as statistics, X-ray methods or controlled experimentation) and no major scientific fields (such as biology, chemistry, nuclear physics and computer science) have been entirely discarded. And it is difficult for us to imagine that they could be. Rather, we cumulatively extend them over time. Our methods and fields of science encompass our extensive bodies of empirical and theoretical knowledge consolidated over time. In contrast, a hypothesis or theory put forth by a scientist can be tested by others and we abandon many of our ideas and hypotheses and a number of our theories, as depicted in Figure 8.1. In analysing over 750 major discoveries, including all Nobel-Prize-winning discoveries, only 1% of discoveries (8 in total) have been abandoned and they were mostly theoretical in nature—not empirical or methodological.[(64)] They were not grounded in rigorous empirical evidence.

It is in particular our central methods of science that are cumulative. Mathematics, RCTs, chromatography, statistics, centrifuges, particle detectors, telescopes, electrophoresis, spectrometers and the like have all been constantly extended and fine-tuned, often over centuries. In contrast to hypotheses and theories, we have not abandoned any of these established scientific methods or instruments. We use them in multiple scientific fields and we continue to improve all of them, which have enabled making dozens of new discoveries. Our methods and instruments make science highly cumulative. In contrast to Kuhn's hypothesis, the history of science tells a cumulative and unified story, and the story is driven by the methods and instruments we develop. Shifting our attention from individual hypotheses and select theoretical discoveries to all major scientific discoveries, methods and fields is the best way to measure and assess the cumulative nature of science. For they make up the foundation of science and how we conduct science and they encompass our established bodies of knowledge. Cumulative knowledge is thus commonly on a spectrum: from unestablished ideas and hypotheses, and then experimental findings and theories (experimental and theoretical discoveries), and finally to established methods and fields (Figure 8.1).

Our best light microscopes today do not compete as a different paradigm to the first light microscope developed in 1590. Our methods of arithmetic

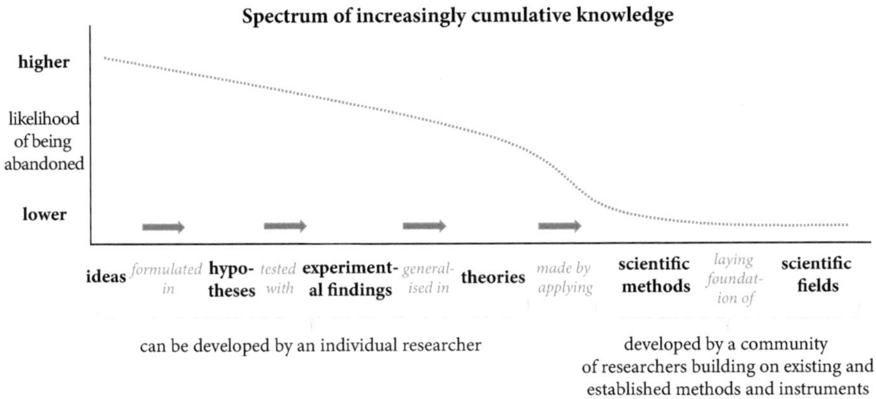

Figure 8.1 The cumulative nature of science: decreasing likelihood of abandoning ideas, hypotheses, experimental findings, theories, methods and fields.

today do not compete as a different paradigm with those developed by the Sumerians. As a vast cumulative project, we have extended our major tools of science over time—from statistics to microscopy. No major scientific method or instrument that is used across fields goes through competing methodological paradigms. They are the foundation of science and being able to do science across fields (so that a possible paradigm shift that would abandon them would so fundamentally change how we conceive science that we could likely no longer call what we do science).

Within the *new-methods-drive-science* framework presented here, scientific progress is viewed as primarily driven by the new methods we develop that enable us to do science in new ways and make new discoveries—not just by largely theoretical developments as Kuhn(1) argues. The Kuhnian focus on a theoretical shift reflects a final output but not the methodological process we take to develop, refine and replicate the output. This new-methods-drive-science framework better reflects actual scientific practice and enables us to better analyse and explain how scientific progress and new discoveries arise. It provides a more coherent alternative to the notion of broad Kuhnian paradigm shifts. While Kuhn provides insight into scientific change in theories, such change in theories is not just historical and intrinsic to science, as he outlines.(1) It is also influenced by a range of other factors, such as cognitive and methodological factors (Chapters 2–15).

The title of Kuhn's best-selling book *The Structure of Scientific Revolutions* portrays the notion of vast scientific changes resembling vast political changes or revolutions (Figure 8.2).[1] A more accurate title, given his specific focus, could have been *An Account of Theory Change Mainly in Early Physics*, as he did not identify an underlying structure across such shifts and they are generally slow, collective processes rather than revolutions. A more accurate account of science could be described as *An Account of Cumulative Scientific Progress Embodied in Scientific Methods and Fields.*

To better understand scientific theories and how we develop them cumulatively—such as Darwin's theory of evolution and Einstein's theory

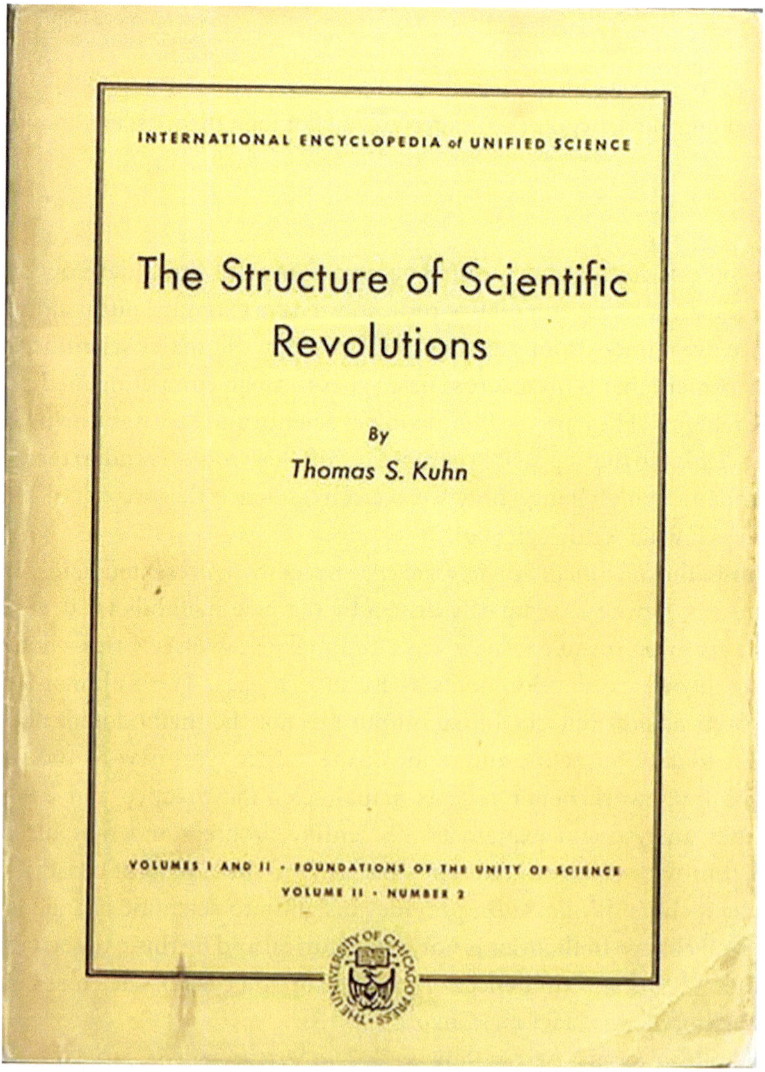

Figure 8.2 Kuhn's seminal book is a foundational text in the science of science, 1962.

of special relativity—we cannot view them independent of their historical context and the existing knowledge on which most of their work builds. Our theories are, when contextualised, small contributions within an existing web of cumulative knowledge. Arguments in favour of evolution, that an organism can descend from another organism, have been made at least since antiquity. They were made by ancient Chinese (like Zhuang Zhou) and pre-Socratic Greeks (like Empedocles and Anaximander),(147) and Romans (like Lucretius) and medieval Arabs (like Al-Jāḥiẓ and Nasīr al-Dīn Tūsī).(148) In 1748, De Maillet also argued that different species emerge by life forms adapting to different circumstances.(149) Erasmus Darwin—Charles's grandfather—viewed living beings as self-improving, arguing in 1794 against a mechanistic view of human life and theorising about evolution that could occur through natural selection.(150) Lamarck developed a similar evolutionary theory in which animals adapt to their local environment and thus use and stop using certain traits. He proposed this as the reason why moles became blind.

In the first few decades of the 1800s, different scholars such as Wells, Matthew, Blyth and Chambers also already developed a theory of natural selection in one form or another. Around the mid-19th century, theories of transmutation were widely accepted among scholars.(149) Then, by drawing on the work of Erasmus Darwin, Lamarck, Chambers and others, Wallace also developed a theory of natural selection in two papers published in 1855 and 1858. He then published work together with Charles Darwin in 1858—before Charles published his expanded ideas in *On the Origin of Species* in 1859.(51) Other scholars had thus already developed the notion of evolution and natural selection before Charles Darwin observed on the Galápagos Islands how mockingbirds varied from island to island, which provided him further evidence of the variation of species.(151) His critical contribution was collecting a range of detailed observations and specimens to support the theory, realising that theories are best grounded with detailed documentation.(149) If Charles Darwin had not written *On the Origin of Species* and remained a hobby geologist, we would still have a well-developed theory of evolution by natural selection.

Einstein is similarly widely credited for developing the famous equation $E = mc^2$, associated with his theory of special relativity in 1905. The simple algebraic equation states that energy E within a system (an atom, the human body or a planetary system) can be calculated as its mass m multiplied by the speed of light c squared. Special relativity explains how speed affects mass, time and space and it transformed our understanding of the relationship between space and time. Yet before Einstein, others already developed variants of this mass-energy equation, and even the very same equation.[152,153] Umov in 1873 and Preston in 1875 developed similar mass-energy formulas, such as $E = kmc^2$, as did Thomson in 1881, Heaviside in 1889 and Poincaré in 1900.[154] In 1904, Hasenöhrl formulated the equation $E = 3/8mc^2$ that he arrived at by studying the properties of blackbody radiation within a moving cavity. And his work was consistent with special relativity.[152] In a paper published in 1903 within the Proceedings of the Royal Veneto Institute of Science, De Pretto developed the same formula $E = mc^2$ that later brought Einstein international fame.[153] This work was all published before Einstein's paper, containing the same formula, was published in 1905 in the German-based *Annalen der Physik*—in which Hasenöhrl also published his work the previous year (which Einstein did not cite).[152] Einstein did later give credit in 1906 to Poincaré's mass-energy equivalence and acknowledged that their formulas were mathematically equivalent regarding the centre of gravity problem.[155]

In 1907, Planck formulated, by building on the existing work of Hasenöhrl and Poincaré, the final derivation of $E = mc^2$. In that same year, Einstein stated that 'It seems to me to be in the nature of things that other authors might have already elucidated part of what I am going to say. However, bearing in mind that the problems under consideration are being treated here from a new standpoint, I felt that I should be permitted to forgo a survey of the literature (which would have been very troublesome for me).'[156] Einstein connected the formula with the theory of relativity, but without Einstein we would still have a theory of relativity which other scientists, like Hendrik Lorentz, were working on at the time. And special relativity was originally called the Lorentz–Einstein theory. Overall, other researchers generally also deserve credit for their contributions to developing nearly all leading scientific theories, from the equation $E = mc^2$ by Hasenöhrl, De Pretto, Einstein and others, to the theory of evolution by Lamarck, Wallace, Darwin and others. It is however easier for science textbooks, teaching science and awarding prizes to associate a discovery with a single name rather than with the community of scholars who developed it (Chapter 6). Imagine if we would have to remember all these names instead of just Darwin and Einstein. But simplicity comes at a cost: it distorts the image of science, how the scientific process actually works and how we foster new advances.

What any one individual can contribute, when we historicise their work, is a piece or connection between already existing pieces, compared to the extensive knowledge and methods developed before them on which they build and make their work possible. Science is, on the whole, a cumulative and iterative process of continual refinement. Darwin, Einstein and the like are not scientific geniuses but people endowed with the same senses and motor skills as others, while generally very good at synthesising and building on existing knowledge and methods. Often they are just smart normal people in the right place at the right time.

In sum, science is cumulative and scientific theories and methods are not independent of their historical context. They are provisional and have been expanded by new evidence, experiments and methodological advances over time. There is a history of science—not just one science that is constant over time. Yet people in the future are not likely to think the same way about the level of validity and rigour of our current theories as we do about those a few centuries ago. This is because today's best methods and instruments (including electron microscopes, X-ray methods and radar telescopes) and the theories developed using them are much better able to explain and predict phenomena with much greater accuracy. Historians of science do not just develop grand accounts of scientific progress, like Kuhn, but also investigate historical documents and correspondences, research science in particular historical contexts and conduct case studies and field work. Interestingly, Kuhn stated that 'in the early stages of the development of any science different men confronting the same range of phenomena ... describe and interpret them in different ways,' and this is evident in the field of science of science.[(1)] History of science helps us understand the highly cumulative nature and evolution of our knowledge and methodological toolbox, and thus the highly cumulative nature of how we build our tower of science. History of science forms an essential piece of the greater picture of science of science—which is linked to the anthropology of science.

9

Anthropology of Science

Anthropology of science is the cross-cultural study of humanity which retraces how we have developed science, from the past to the present. So how did we get from being hunters and gatherers using axes, basic observation and numerical reasoning, to being scientists hunting explanations and gathering data using systematic controls and mathematical methods? How did we get from being nomads running in the savannahs and conducting trial and error, to being scientists running sophisticated experiments and conducting statistical analysis? We do not have a definitive answer as we face constraints to reconstructing past events. This holistic and methods-driven account here however provides an integrated explanation.

Our species began to observe, solve problems, experiment and acquire knowledge more and more systematically, on the one hand, as our biological, cognitive and social abilities evolved over time by interacting with our environment (Chapters 2–5). On the other hand, we did so as broader social, economic and historical factors enabled us to use these abilities more effectively and build on what we observe in larger groups (Chapters 6–8). We developed more and more complex language an estimated 50,000–100,000 years ago.(92) Language is essential to be able to express our observations and ideas verbally and explain them to others—both then and today (Chapter 15). Expanding our language abilities allowed, in an interconnected way, for greater cooperation in larger groups and greater tool-making. This made it easier to pass along the methods and tools we developed and the knowledge we acquired about flora and fauna. We became increasingly better botanists, zoologists, geographers and engineers within our particular environment. Together, these expanding cognitive and social abilities gave us an increasing advantage, likely for the first time in history, over many other smart animals—animals that also have such abilities, though to a lesser extent (Chapters 2–4).

We then created basic measurement tools such as simple tally mark systems at least 35,000 years ago. Eventually, by developing agriculture (an estimated 12,000–13,000 years ago)(92) we increased our availability of food and our labour productivity. This freed up our time and cognitive resources as we gradually abandoned our hunter-gatherer lifestyle to become sedentary village

Science of Science. Alexander Krauss, Oxford University Press. © Alexander Krauss (2024).
DOI: 10.1093/9780198937401.003.0010

dwellers. We could increasingly dedicate our time to other cognitive activities, beyond meeting our basic needs such as food, shelter and clothing. We created a wider range of goods and technologies and experimented with more productive crops. We eventually developed written language and complex numerical systems—two essential features of contemporary science—an estimated 5000–6000 years ago.[92] These incredible new tools enabled us to document, calculate and plan food production at a mass scale for growing populations. With written language, we could then record what we observed and experienced and better share it across generations, providing an external source of memory (without forgetting) (Chapter 15). At the time, to be able to develop technologies and tools, we used trial and error, other forms of basic experimentation and tested hypotheses—though often implicitly. As villages expanded into cities and some cities into empires over the past several thousand years, growing populations were able to build on cumulative knowledge and increasingly specialise.[92] We could then develop more sophisticated methods, tools and knowledge (Chapter 7).

Some cultures maintained a degree of stability over centuries, and even millennia, such as the ancient Chinese, Greeks and others.[57] Such stability allowed us to build extensively on what we know and to develop a wider range of scientific and technological tools. With more complex numerical systems, we became able to record and measure what we observed more systematically—from agricultural plots to the movements of the moon. Population density, specialisation (division of labour) and methodological diversity increase together, with changes in one generally affecting the others. Together, they have allowed us to grow up in a cultural context with much cumulative knowledge about the world (Chapter 7). Scholars especially in ancient China and Greece studied a wider range of phenomena than in earlier civilisations, from astronomical events and the properties of living animals, to magnetism and sound. They did so with a logical view of how the world is broadly construed and they viewed certain phenomena as operating according to general principles.[57,157]

Adopting a pragmatic experimental approach, ancient Chinese developed, as the first or independently, many more advancements than their Mediterranean counterparts, the ancient Greeks.[157] These include effective immunisation techniques, magnetic compasses, negative numbers and the 'Pascal' triangle, astronomical observations of novae, seismographs, paddlewheel boats, irrigation systems and quantitative cartography, as well as papermaking and printing that fostered the spread of knowledge.[57,157,158] Ancient Chinese created smallpox vaccines, which required a complex understanding of the causes and effects of infectious disease, their interactions and how to control them.[157] It is because the Chinese created a more complex system of astronomical records than any other culture—including star catalogues and observations of eclipses and novae—that our records today are able to go back millennia.[94,157] Such advancements allowed ancient Chinese to control and predict nature through technology.[57,159] Not only medicine but also geology, alchemy/chemistry, geography, technology and engineering were bodies of knowledge supported by the Chinese state.[94]

In the centuries leading to the 1500s and 1600s, we began to exchange technologies more rapidly and eventually for the first time globally.[92] We had already widely used the methodological approaches of more systematic observation, measurement and experimentation to create increasingly sophisticated technologies. These include eyeglasses, windmills and mechanical clocks in 13th-century Europe (though mechanical clocks were already created in 8th-century China) and the microscope in the 16th century.[160] Eventually, these methods and instruments were applied to questions whose practical relevance was not only directly observable (technological knowledge) but also increasingly not always directly observable (purely scientific knowledge). At the time, we increased diversification and productivity at a scale not yet experienced in history.

What made the work of 17th-century scholars possible is a cumulative process of greater technological advances and greater awareness of these more systematic methods already widely used for such advancements. These methods were then adopted to also study more theoretical questions about the world, commonly using the newly developed instruments including microscopes, barometers and telescopes that made many of the discoveries possible. At the time, scholars expanded science by combining our evolved methodological abilities and adopting written language, mathematical systems including geometry and algebra, and diverse technological and scientific knowledge developed by our ancestors over thousands of years. They built on the existing work of scholars like Aristotle, Archimedes and Ibn Al-Haytham. The political environment in parts of Europe then became increasingly open to question the status quo and authority.

The printing press, first developed in China and later brought to Europe, supported the spread of ideas and freedom of thought.[157] Scholars like Copernicus, Galileo, Boyle and Newton were thus able to grow up in a context with much cumulative knowledge and favourable demographic, economic, technological and environmental conditions. They grew up in a context rich in natural resources and with productive agricultural practices that supported the necessary surplus of labour that enabled them to focus on studying the world.[(92)] Using methods in more systematic ways became institutionalised through newly established education systems as centralised governments spread in the 19th and 20th centuries. These became our main means to systematically transfer knowledge and complex methods across generations. Since the second half of the 20th century, developing digital technology and computers has exponentially accelerated our ability to produce and distribute knowledge. Today, we continue making extensions to our methodological toolbox that allows us to acquire increasingly sophisticated bodies of knowledge (Chapter 10).

These are, taken together, the historical hallmarks in the development of our human mind, social organisation and methods and instruments that have enabled us, in increasingly complex ways, to develop science. This account outlines how we over time developed our mind's methodological abilities and more complex abilities for language and a numerical and measurement system, and eventually we became more systematic in observing, experimenting and creating theories about the world, and ultimately began gathering large sets of data observations and creating complex methods such as statistics to analyse them. The combination of these multiple abilities and factors, and especially the methods we developed, gave rise to growing communities in which we more and more systematically experimented with and developed better hunting techniques, then better agricultural crops, eventually better plant-based medicines and more complex technology, and ultimately sophisticated theories about the origins of life and the universe. It is an account of the history of human reasoning, knowledge and science, summarised in this chapter. The evolution of our knowledge is marked by continual progress when viewed over long historical periods, though it is not entirely linear. Religious dogma, famine and war have led to knowledge stagnating at different times throughout history, such as in parts of the Middle Ages.

Also, differences arise across cultures in the methodological approaches we develop using our universal methodological abilities, and thus in the ways we think and our world views. These are shaped by the particular environment we are exposed to.[(57)] For ancient Chinese, thinking can be characterised as being more holistic, continuous and observation-based, and ancient Greek thinking as more reductionist, principle-searching and abstract.[(57)] Ancient Chinese, by adopting a holistic and experimental approach to reasoning, developed a vast range of systematic methods and remarkable technologies as outlined above.[(157,158)] The common view is that scholars began to systematically study and understand the world only around the 17th century in

Europe.[121] But that view neglects the technological and scientific advances and knowledge made using systematic observation and experimentation by the ancient Chinese, Arabs and others that Europeans built on.[82]

Ethnographic research of indigenous groups worldwide describes how they have also acquired comprehensive knowledge, over generations, about our natural environment including plants and animals, their characteristics, properties and lifecycles and developed sophisticated classification systems for them.[55,56] An estimated 70% of medical drugs are plant based and much of what we know in science about the healing potential of plants has been acquired from indigenous cultures.[161] From these cultures, contemporary science has adopted much experimental knowledge about nutrition, botany, agriculture, natural pharmaceutics, ecology and agroforestry (Chapter 3).

Now consider a group of people who do not commonly develop abstract theories about reality and who are viewed, by contemporary scientists, as being constrained in acquiring reliable knowledge. They use sensory and reasoning abilities including observation and experimentation, such as trial and error, as the main means to solve problems and acquire knowledge about the world. And they are thus not familiar for example with the theory of relativity or quantum mechanics. Consider also another group of people who, in addition, create and use an astronomical map for mathematical and observational purposes, and apply a 0 to 60 scale as the basis of their numeral system. Both of these groups have rich knowledge of plant-based medicines, the characteristics and behaviour of animals and many other features of the world. They understand buoyancy and construct boats. They can think in mathematical terms. But for contemporary scientists, their methods may—despite developing much practical knowledge—be viewed as rather primitive. The first group refers to hunter-gatherers, and the second group to Sumerians.

In sum, science has developed as a gradual process over our species' history, in which we have used our evolved methodological abilities of the mind increasingly systematically and collectively. Over time we have strengthened the foundation and structure of our tower of science. We have gained, used and further developed extensive knowledge and methods from our earlier ancestors. These have been absorbed into contemporary science—from the plant-based medicines of indigenous populations to the technological, scientific and mathematical advances of ancient cultures. Science, commonly defined as the study of the 'world through observation, experimentation, and the testing of theories,'[65] is not just a product of 17th-century Europe. Only when we integrate evidence about human evolution and the mind

(Chapters 2–5) with social organisation (Chapter 6) can we gain a more coherent understanding about the foundations of how we acquire knowledge and do science—and how we expand science, when we also integrate evidence from the broader range of factors (Chapters 7–15). Anthropology of science is just one integral part of understanding the foundations of science. It helps us explain how we have gotten to this point in history and provides much evidence of how the methodological roots of science have evolved.

10
Methodology of Science

Over our species' history, we evolved methodological abilities of the mind (observation, problem solving and experimentation) that we use together with increasingly complex methods developed using these abilities (controlled experimentation, statistics and X-ray methods). Science has always been grounded in these evolved methodological abilities (our *universal* methodological toolbox) that have enabled us to develop vast bodies of knowledge by creating sophisticated methods and tools (our *adaptive* methodological toolbox). As we face constraints when using our evolved abilities to do science, we have developed methods and instruments to reduce these constraints. Such constraints are cognitive (such as limited sensory abilities, cognitive bandwidth and memory), social (such as cultural values, norms and interests), geographic (such as differences across contexts that require conducting studies in multiple contexts) and so forth (Chapters 2–8 and 13).

A central argument throughout the book is that we develop new methods and instruments that enable us to better access and understand the world and make new scientific advances by addressing our human and methodological constraints (Chapter 4). Mathematical methods for example are used across fields, from theoretical physics to economics, to help us systematically calculate and measure phenomena and represent them using algebraic equations (Chapter 13). Controlled experimentation and randomisation (such as randomly allocating participants to groups in an experiment) are used across fields, from biomedicine to psychology, to reduce human biases in designing, implementing and analysing studies. Datasets allow us to store information and sets of observations externally. Magnetic resonance imaging (MRI) enables us to detect phenomena like magnetic fields and radio waves that we do not have sensory receptors for and is used widely for medical purposes. Computers aid us in efficiently processing, analysing and sharing large sets of observations in ways that would not be possible without them (Chapter 12). Electron microscopes vastly enhance our visual capability and enable us to perceive miniscule objects using the wavelength of an electron, far exceeding the magnification of light microscopes. Nanoparticles, microorganisms,

Science of Science. Alexander Krauss, Oxford University Press. © Alexander Krauss (2024).
DOI: 10.1093/9780198937401.003.0011

crystals and molecules come into our visual scope with an electron microscope (Figure 10.1).

Such methods and instruments we create using our mind greatly extend our evolved mind. Our methods and instruments, once created, are mainly external resources that can function as an efficient external extension of our mind (Chapter 4). In general, when we design any new method or instrument, we are asking how we can answer a question by improving our existing cognitive, sensory or methodological abilities. We are asking how we can address a problem by reducing a human constraint or bias to doing science.

A companion study analysing the methods and instruments used to make over 750 major scientific discoveries, including all Nobel-Prize-winning discoveries, provides direct evidence of how we drive science. It illustrates that 94% of discoveries used observation, 75% experimentation and 100% developed a new method or instrument (X-ray methods, radar telescope etc.) that was applied to make the discoveries.[64] The main conclusion is that scientific methods and instruments we develop drive new discoveries—that is, they are the main mechanism through which

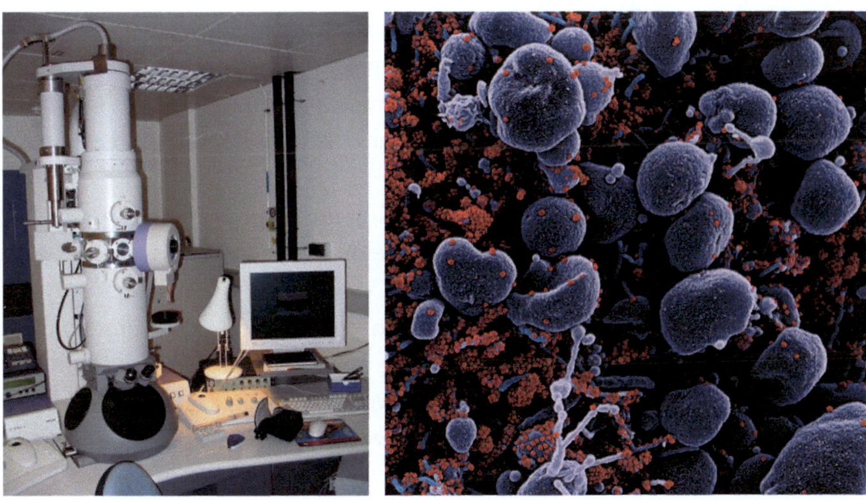

Figure 10.1 An electron microscope, one of the most important scientific instruments used across fields (left), with a photograph of coronavirus taken using it (right).
Source: David Morgan; NIAID.

we have directly made scientific discoveries. The study has five main findings:

- Directly after developing a method like modern statistics and X-ray diffraction, or an instrument like the electron microscope and chromatography, we have been able to make dozens of scientific discoveries with each of these methods and instruments.
- Discoveries can only ever be made after we develop the needed methods or instruments and are commonly made soon after developing them: 10% of discoveries have been made simultaneously or in the same year of creating the new method or instrument used to make the discovery, 20% within two years, 32% within four years and 52% within ten years.
- The ten big methods and instruments that have been most commonly used to make discoveries and have revolutionised science are statistical/mathematical methods, spectrometers, (electron) microscopes, X-ray methods, chromatography, centrifuges, electrophoresis, telescopes, thermometers and lasers.
- There are three ways new methods and instruments drive new breakthroughs. One way is that a researcher develops a method or instrument themselves that they apply to make a scientific discovery (25% of all discoveries). The second way is that a researcher develops a new method or instrument that is applied by another researcher to make a scientific discovery (50%). The third way is that a researcher develops a new method or instrument that is the major discovery itself (25%), such as the electron microscope and particle detector.
- Many new scientific tools enable making multiple discoveries, at present and in the future, and in different fields beyond which the tools were designed. They thus causally enable discoveries unforeseeable by the inventors of the tools. The discoveries of cells, bacteria and mitochondria, and Uranus, galaxies and pulsars were only possible after developing the microscope and telescope and they were not being searched for. Many breakthroughs are exploratory (not hypothesis-driven) that we make by applying a new scientific tool.(64)

Scientific advances are thus made possible through new methodological innovations. While conventional research applies existing methods, cutting-edge research that produces new discoveries uses newly developed methods or instruments that enable studying the world in a new way. This direct link between new methodological innovations and new discoveries is observed in Figures 10.2 and 10.3. And this direct link has been largely overlooked in understanding what drives science and its limits.[(64)]

New scientific methods and instruments drive not only new scientific discoveries but also new scientific fields. Developing particle detectors and accelerators gave birth to high-energy physics/particle physics in the early 1900s, and the field continually expanded by developing more advanced detectors and accelerators. Creating the electron microscope in 1933 helped open the field of molecular genetics. Developing X-ray diffraction methods in 1912 gave rise to protein crystallography in 1946. Femtosecond spectroscopy created in 1985 led to the emergence of femtochemistry in 1988 (Figure 10.4). Newly developed methods and instruments are often the defining feature of new fields.[(64)]

Another dimension of scientific methods is that each method uses a common measurement, and a common measurement is closely linked to being able to develop more replicable and reliable knowledge among independent scientists. Rigorous findings can then be viewed independent of individual scientists, as they can be tested for accuracy by a group of scientists using

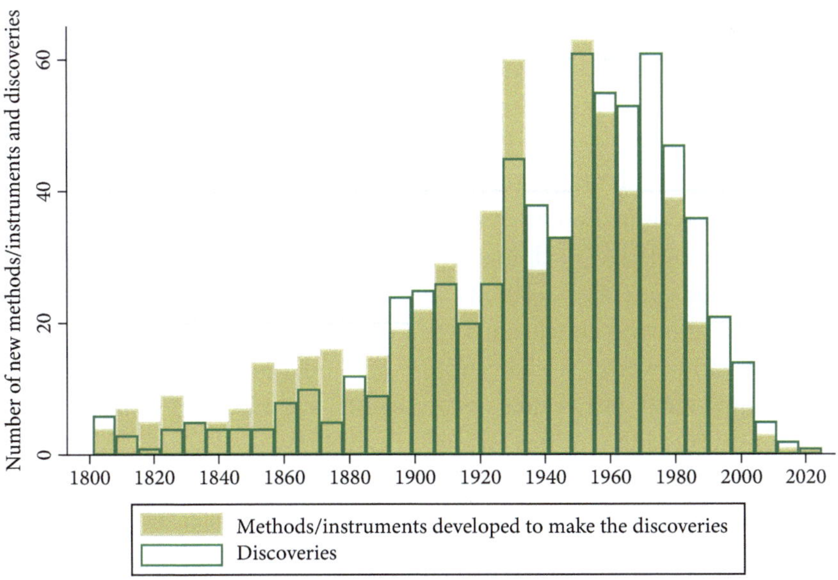

Figure 10.2 Trends in the development of new methods and instruments closely follow trends in subsequent discoveries.

Data reflect the number of central methods/instruments and discoveries made in a given period among 653 major discoveries, from 1800 to 2022, and include all Nobel-Prize discoveries.[(64)]

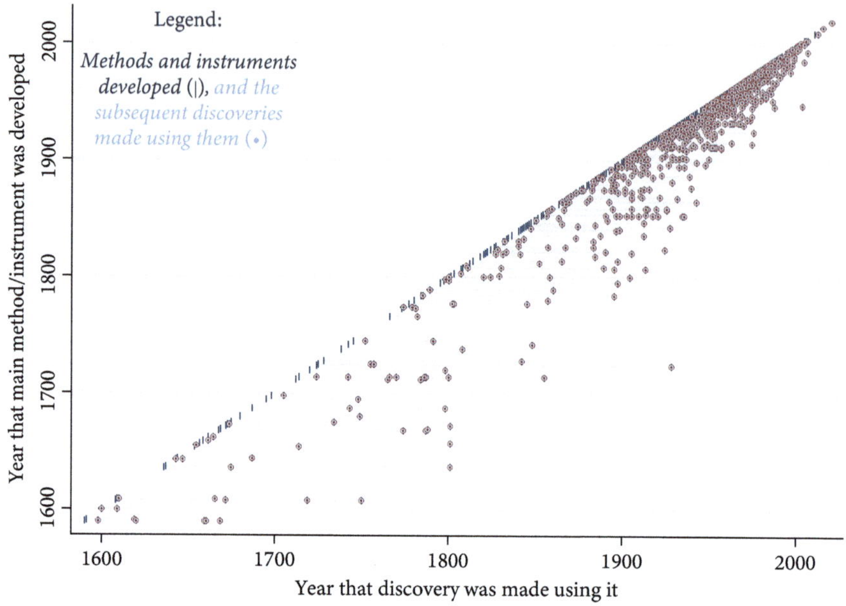

Figure 10.3 New discoveries are sparked by developing methods and instruments needed to make the discoveries.

Data reflect 734 major discoveries since 1575, including all Nobel-Prize discoveries, and the central methods/instruments used to make them. Data illustrate that the year the main method or instrument is developed is strongly correlated with the year the discovery is made.[64]

the same measurement method. This reduces the possibility of individual scientists influencing results (Chapter 13).

However, methods and instruments, while reducing human influences and enhancing our limited abilities, can bring constraints and possible biases in developing knowledge. Each mathematical technique, X-ray method and statistical method generally has limits as to which questions we are able to study and what results and conclusions we are able to derive using it. Each has a set scope within which we can capture or model phenomena in the world, design, implement and evaluate experiments, and interpret results.[42] Each faces constraints and requires making assumptions.

One of the best ways for us to reduce individual methodological constraints in science is by applying multiple methods.[40,42] For each method can provide different evidence and perspectives into a phenomenon. Each method can confirm whether results consistently point in the same direction. To better understand for example how a medical treatment may affect patients or how a government policy may influence citizens, it is best to apply multiple methods. These range from quantitative methods including RCTs and observational

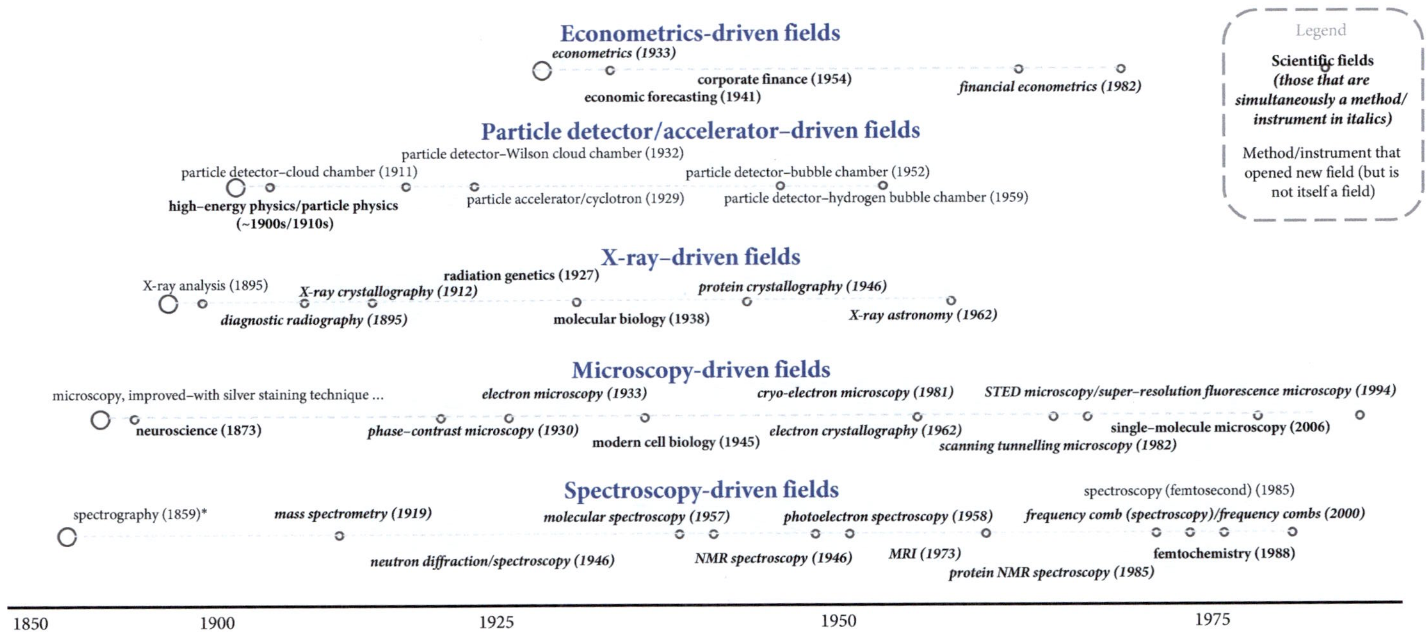

Figure 10.4 Developing a new central method or instrument opens new scientific fields (illustrations from Nobel Prizes for developing methods/instruments and subsequent fields).

*The first spectrograph, developed in 1859, is the only instrument included that did not receive a Nobel Prize, which was used to open the field of mass spectrometry.[64] NMR, nuclear magnetic resonance; STED, stimulated emission depletion microscopy.

studies, to qualitative methods including a set of rich single cases, consensus among a group of experts and historical evidence.(ibid.) Methodological constraints can at times also be reduced through statistical simulations, bootstrapping (a resampling method in statistics) and calibration. Yet assumptions, limitations and biases that can influence our results and knowledge cannot always be entirely avoided. Some form the basis of our very methods (discussed in Chapter 13).(ibid.)(162) Others enter the scientific process before and after we apply methods and thus cannot be reduced using methods. These range from data collection techniques and selecting study design features, to interpreting outcomes and using the results beyond the study context.(42)

At a deeper epistemological level, with the range of methods and instruments we have developed thus far in history, we do not have an inherent inclination to view the world using one particular method or classification system instead of another. We are not naturally inclined to use the metric instead of imperial system, Fahrenheit instead of Celsius, probabilistic instead of non-probabilistic methods, a numerical system with 10 as its base instead of 60 and quantitative instead of qualitative methods for all questions (Chapters 8 and 9). We do not have an innate predisposition to acquire knowledge and conceptualise phenomena in the world using mathematical equations and universal laws instead of using no mathematics and no laws. Which method we develop and use depends on our particular question and scientific field. We do not inevitably explain phenomena using one disciplinary perspective or method instead of multiple. But it is how we are methodologically trained and which methods we have developed thus far that leads us to do so (Chapters 2–6). Our scientific methods such as computational and statistical techniques and X-ray methods that we recently developed, and thus our theories we create using them, are in a process of continual refinement over time.

In sum, because our evolved mind faces constraints in perceiving, processing and explaining our complex world (Chapters 4 and 5), we have developed a powerful methodological toolbox that significantly expands our cognitive abilities and bodies of knowledge. Our tools of science in turn can face constraints in how we represent and model the dynamic character of phenomena using them. Our mind's methodological abilities and complex methods and instruments we create using these abilities (our universal and adaptive methodological toolbox) are at the centre of understanding science. Methods and instruments we develop are a factor we can directly influence and are very important in shaping the foundations, limits and advancement of science, as we illustrate comparatively to other factors in Chapter 16. Our scientific methods and instruments are also the only factor that underpins all other factors

(Chapters 2–15). A central focus on scientific methods is thus essential to the integration of science of science. Our *tower of science* is made possible by our evolved methodological abilities (that account for its foundation) and our scientific methods and instruments (that account for the different floors of its structure), which together determine how far we can observe, measure and understand the world. Ultimately, the discovery of the methods and instruments we apply are rarely cited and referenced in our studies even though they make our studies and advances possible. So citations do not capture their enormous impact on science and this is a shortcoming in our scientometric system.

11
Scientometrics and Network Science

Science describes and explains the world through research articles and books that are organised into scientific fields. Scientometricians including network scientists analyse this scientific literature. To do so, they rely on the indicators of citations and publication counts to study issues such as research productivity, team collaborations, career dynamics, networks of scientists and institutions, and novelty in science.[4,5,8,9,35,163,164,38] They use large-scale data (big data) and network analysis and search for patterns in such data. That is scientometrics in a nutshell. Scientometrics is of interest to all researchers as citations and publication counts largely determine whether researchers get research grants, academic jobs and promotions. Most measures of scientific impact and success use citations (the number of times a publication is cited by other researchers). Our current reward system in science is deeply embedded in this metric.

The British bibliometrician and historian of science Derek John de Solla Price is commonly considered the father of scientometrics. He pioneered work in the quantitative analysis of scientific literature and citation analysis to assess scientific productivity. In his book *Little Science, Big Science* published in 1963, de Solla Price examined the structure and growth dynamics of scientific knowledge.[70]

Since then, a central topic in scientometrics, and in science of science in general, has been innovation. Studies illustrate that researchers are generally risk-averse, choosing to study phenomena in which they already have expertise and with which they are familiar. This limits what is studied in the future and making potential new discoveries. Researchers willing to explore new areas and undertake a riskier career, moving from traditional topics to riskier innovation, are more likely to expand a field and make new discoveries. Research strategies that are conservative foster individual careers rather than science as a whole. What characterises high-impact science are conventional combinations of existing work that integrates novel combinations of not-yet-connected topics[165,166,167] or research methods. To increase the impact of innovative and cross-disciplinary research, scientists need to show how it contributes to established research. Otherwise research is less likely to achieve

Science of Science. Alexander Krauss, Oxford University Press. © Alexander Krauss (2024).
DOI: 10.1093/9780198937401.003.0012

its maximum impact.[166] When it comes to incentivising innovation, a study finds that medical researchers receiving greater freedom in experimentation and failure and greater rewards for long-term success are more likely to produce higher impact research.[168] Funding bodies can thus strategically foster higher risk research projects that test unexplored ideas, areas and diseases[5] and unexplored research methods. Funding bodies and academic journals also evaluate submitted manuscripts and project proposals partly by predicting their potential future impact.[35]

In terms of researcher productivity and impact, major discoveries are generally made by younger researchers and explained by their higher productivity and not yet securing permanent positions.[4,7] Studies generally illustrate a median age of discoverers between their mid-30s and mid-40s.[4,7,169,170] Scientists of all ages can make major discoveries as long as they are productive, though the likelihood decreases over scientists' careers.[7] Assessing all Nobel-Prize-winning discoveries shows that only 7% of discoveries were made by scientists after the age of 50 and only 1% after the age of 60.[64] So if researchers do not make their great contribution to science earlier in their career, they are unlikely to do so and receive a Nobel Prize in their lifetime.

As science expands, each generation has a larger body of knowledge and methods at its disposal. This generally leads to more specialised (and at times longer) education and greater teamwork in science (Chapters 6 and 7).[38,171,119] So reaching the research frontier takes longer and longer.[8] A study of about 20 million research articles reflects a related trend of a general shift towards teams across all scientific fields: from slightly over one to about five individuals per team in science and engineering between 1900 and the early 2000s.[172,173] Larger teams can often be better able to develop new combinations of ideas and apply different methods (Chapters 6 and 7). Greater collaboration also generally increases research quality and impact, with prominent European researchers collaborating with other prominent European researchers more often than prominent North American researchers collaborate with each other.[174]

Another central question in scientometrics, and in science of science in general, has been what drives the world's top researchers.[4,5,7,35,164] A study of early-career factors driving success in science identifies four key factors among the world's top 100 researchers across fields (measured by the highest h-index): co-authoring with other top 100 researchers, working at a top 25 ranked university, publishing a paper in a top 5 ranked journal and publishing most papers in first quartile (high-impact) journals. Over 95% of the 100 researchers with the highest h-index, across multiple fields, had at least one of these four features in the first five years of their career.[175] Co-authoring with leading researchers is one of the best ways to become visible early on in a scientist's career, through more citations and mentorship.[176] Being at one of the top 25 ranked universities in one's early career is related to greater research impact later, as researchers can enjoy a high-quality research environment and access to greater resources. Researchers' early career is thus important, with the choices and effort that young researchers make shaping their later success (Figure 11.1). Yet because these early-career attributes of successful scientists are predictable of later success, it suggests that the scientific system is already relatively closed. It also points to shortcomings in using the common and highly influential indicators of success, namely citation metrics. This is because early career advantages—measured using these metrics—are so strong that they help predefine 'highly successful scientists' without further information about the content or social and policy impact of their research.[175]

Measuring scientific success using citation counts can however constrain us in developing new ideas since it gives an advantage for highly-cited researchers to become even more cited—that is, the rich just get richer. In turn, it provides a bias for researchers, with limited time, to use more cited and thus at times older research. It also disadvantages younger researchers and innovative researchers working between disciplines and paradigms.[35,131] Scientific institutions need to place greater focus on other metrics of success beyond citations, such as levels of innovation and societal relevance of research[5] and the development of new methods and instruments.

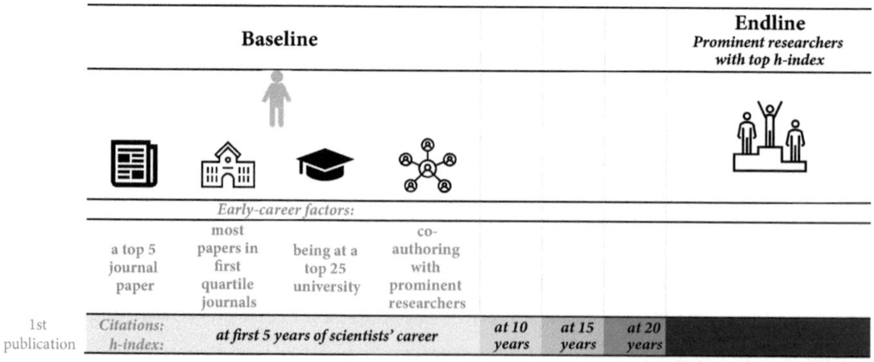

Figure 11.1 Early-career choices and factors can shape later success in science.[175]

Tracing science and the scientific process through published literature and citations (the scientometric approach) does not uncover many relevant factors to understanding science. Citation counts are a metric that does not capture the impact of most major past discoveries that account for our vast bodies of knowledge that make up science, because most major discoveries throughout history were made before the widespread use of references and citations that began in the second half of the 20th century.(177) Citation counts do not capture the immediate impact of new ideas or breakthroughs in science (as citations take time to accrue) or the impact on policy or society (as they cannot be cited).(39) Citation counts do not capture well the powerful role of our scientific methods and instruments in conducting and advancing science. Studies at most mention techniques and procedures used in an experiment but rarely cite and reference the discovery of the method or instrument they use. Our most important scientific methods and tools, such as Martin and Synge's partition chromatography method,(178) Townes' maser/laser,(179) Ruska's electron microscope(180) or Svedberg's centrifuge,(181) would each have received millions of citations as they are mentioned in millions of studies, according to Google Scholar. This far exceeds any scientific studies in any field. But these method-making studies only have received between a few hundred and a few thousand citations each.

We require a new norm in science of always systematically citing the methods and instruments we use, so their vast importance becomes powerfully evident. This book reflects a fundamental rethinking needed in how we study and understand science, moving away from just using ex-post indicators of output, mainly citations and publications,(4,5,6,7,8,35,36) in order to understand all other features of science that they do not capture. It reflects a shift towards also studying the equally important ex-ante indicators of the process and inputs of science and discovery. This requires placing our methods and instruments at the centre of study but also greater focus on the broader demographic, social and economic contexts of science—which are not captured with citations (Chapters 2–15). Also, most existing publications in scientometrics study one factor using descriptive data. This common approach in the field resembles the early emergence of the social sciences before the advent of more rigorous experimental designs, controls, regression analysis and causal identification strategies. Using such methods, we can identify the importance of one factor relative to other factors and can provide causal explanations. Big-data has not been able to push the field forward with robust causal knowledge. For that, we have to turn to more rigorous methods used in fields like public health and economics, and we discuss these in the conclusion.

In sum, studying scientific literature and citations contributes to our understanding of science by providing insight into the conditions shaping scientific impact, collaboration and productivity. Scientometrics alone, like any science of science subfield, cannot however provide us with a comprehensive understanding of science. Scientometric research relies heavily on using computers and statistical methods, which are both used in all scientific fields and are important pillars of the tower of science that we discuss in the next two chapters.

12

Computer Science of Science

We are constrained by our limited cognitive and computing capacity when studying the world. We are flooded with vast amounts of new data and publications each year, at a pace far exceeding our human abilities to process the expanding influx of information and data. Computers play a central role in science and studying science by expanding our limited cognitive resources, memory and capacity for data processing, statistical analysis and simulations (Chapter 4).[44] Computers are used across all research fields, completely transforming the way we do science. Computational methods are also the foundation of a number of scientific fields—from computational physics, biology and cognitive science to data science and bioinformatics. Computers are so central to science that they have become an integral piece of the overall picture of science of science.

New discoveries and methods can now spread across the globe and reach scientists with computers and internet basically instantaneously. We now have almost immediate access to the vast range of existing methods and bodies of knowledge in science, for the first time in history. The World Wide Web is today's version of China's Yongle Encyclopaedia in the 15th century that aimed to bring together all available knowledge in the world in an encyclopaedia.[182] Computers are crucial in conducting science much more efficiently by automating scientific processes and making it much easier to organise, store and retrieve enormous amounts of data through databases. Computers and the internet also connect scientists in ways and in numbers unparalleled before, allowing for much greater collaboration around the world.[44] In short, computers do such a good job at maximising efficiency that we have become highly dependent on them—and it is difficult to imagine our lives without them.

A critical bottleneck in making computers and the internet possible was overcome in a landmark article titled *A mathematical theory of communication*, in 1948. The American mathematician and electrical engineer Claude Shannon published this seminal article in which he addressed the question if there was a unified theory for communication—a general theory for how we produce and transfer information.[183] His answer was to conceive the

Science of Science. Alexander Krauss, Oxford University Press. © Alexander Krauss (2024).
DOI: 10.1093/9780198937401.003.0013

digital nature of information as binary digits (0 or 1). This completely shaped how we began to use computational data. The general model of communication he developed is simple: sent information (data) becomes the received information (data) (that can possibly be distorted by noise)—that is, a transmitter converts information into a signal (that can be distorted by noise) and is then decoded by the receiver (Figure 12.1). Central to the idea is that Shannon modelled the transfer of information probabilistically. Today, data are widely used as binary digits across science, for example in statistical analysis to capture phenomena in the form of variables.[183] By demonstrating how we can quantify digital information, Shannon's work has been called the Magna Carta of the digital age.[184] It helped pave the way for scientific computation, complex quantitative analysis and mathematical modelling. Scientific computation enables simulating phenomena—from physical and biological systems to economic actors and societies. Algorithms embedded in statistical computer programmes (like Python, R and Stata) enable computing complex regression analyses and simulations using thousands or even millions of data observations.

A rapid increase in computing power and available data has also accelerated growth in artificial intelligence, providing new opportunities to apply computer technology in science.[44,185] We most directly advance science by developing new methods and instruments that address our cognitive and methodological constraints (Chapters 4 and 10) and there is great potential for methods of artificial intelligence to expand and automate aspects of the way we do and study science beyond our mind and conventional methods. Machine learning applies computer algorithms that improve automatically through an iterative process of using a given dataset. It allows us to delegate some aspects of data collection and analysis to automated computer programmes.

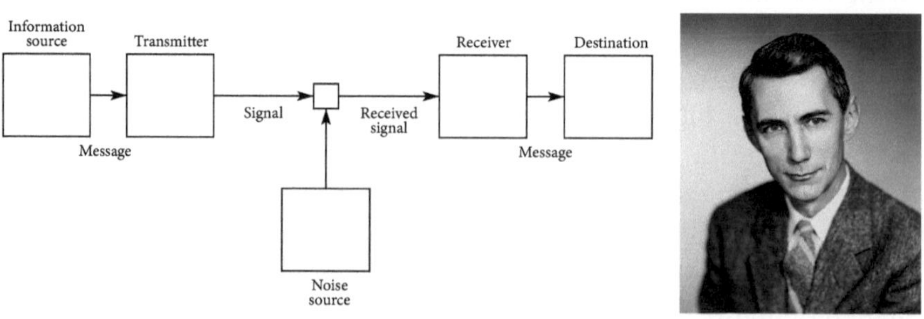

Figure 12.1 Diagram of Claude Shannon's model of communication that made our digital age possible.[183]

Source: photograph from Tekniska Museet via Wikimedia Commons.

This is especially relevant when we study phenomena for which we have large amounts of data or when we require making quick decisions.

In the biomedical sciences for example, we have methods for drug design that automate many mechanical tasks performed by biomedical researchers. Researchers provide the collected data that are coded and inputted into robotic platforms that automatically conduct a series of experiments and generate results.(46) Deep learning methods can help medical practitioners decide whether brain scans and images of patients indicate signs of a stroke or tumours, with algorithms reaching similar accuracy as medical specialists but much more quickly. These new methods can complement (not replace) human expertise.(45,186) Machine learning has also been applied to help predict protein structures using large genomic datasets, to estimate the effects of climate change on cities using large climate datasets and to detect signals in large astronomical datasets.(187) Artificial intelligence can help us solve problems and make decisions when we deal with greater complexity and it can help us reveal new structures and patterns in large amounts of data—data mining.(187) Artificial intelligence programmes are also being designed to rerun data analyses with an existing or new computational method and produce new hypotheses that can verify or diverge from existing research.

Artificial intelligence systems may reduce some biases and errors that arise due to human influence in the scientific process (Chapters 4 and 5),(4,44) although they also give rise to other biases and errors that only manual data collection and analyses can detect and avoid. This is due to the diversity in methods, evidence and language across studies and fields that constrains generalised automatisation using computers. Such programmes can help conduct part of the routine work in science, so researchers could dedicate more attention to other (including creative) aspects of science. Yet we face important constraints in understanding what factors actually lead such programmes to particular results (that is, understanding how they arrive at results) which are thus not always replicable.(4,44,185)

Computer scientists have also offered a computational account of discovery by attempting to model and simulate scientific discovery processes.(188,189,190,191) Computational accounts have been ambitious, attempting to develop computer programmes and algorithms that could drive new discoveries. But they have mostly only focused on the path from data to scientific laws, and do not analyse the role of methods and broader background factors in the discovery process that are taken as given. To date, they have had only limited success in reproducing past scientific discoveries and do so at a high level of abstraction.(187,188)

Greater computational power has led to big data and the analysis of vast complex datasets. This can help us overcome small sample sizes and low statistical power, but it generally brings more noise to data and faces the same challenges that arise with smaller datasets.[(192)] In general, artificial intelligence and machine learning are becoming increasingly important for some tasks of collecting, cleaning and analysing data in different fields. Large language models (like ChatGPT) can assist certain researchers with various tasks, functioning as a kind of research assistant. These include quickly summarising key points and trends in a field and suggesting potential hypotheses and questions to explore based on existing knowledge. They can help with coding, identifying patterns in data and offering preliminary interpretations of results, but can also assist with editing and improving the clarity of text. All outputs must be validated and updated by researchers. This enhances productivity and enables researchers to focus on higher-level tasks. At earlier conceptual stages of research, they seem to be more helpful than at later stages. Overall, these tools are expanding rapidly, with new developments continually being made.

Though, the early hype around new methods like initially with big data analysis and network analysis can fade as we become aware over time of the limitations and biases facing the methods across different fields. Science, despite computational advances and our new dependency on computers, remains human centred. Experimentation and deep methodological, empirical and theoretical understanding are central aspects of science and they remain largely driven by humans, not computational machines. While we delegate more tasks in science to computers and while some fields become more computational, science continues to be led by humans—by our mind, methods and instruments that we are presently better able to understand than many artificial intelligence programmes.

In sum, computers have become an important tool for us to conduct and study science much more efficiently, particularly how we share and use data, methods and knowledge—quickly and globally. By transforming how science is conducted, computers have become an essential part of the foundation of science, and thus our tower of science. We have used artificial intelligence methods in tasks such as processing and analysing data, and classifying and detecting statistical patterns—though they have not yet been able to make new major discoveries alone.[cf. (187)] Using computer technology in science is also closely linked to data science, with scientists using statistical computer programmes and binary digital data (0 or 1) across all fields for data analysis, modelling and predictions.

13
Statistics and Mathematics of Science

Revolutions are easy to spot when they happen as a single extraordinary event. The impact of some methodological discoveries, such as X-ray analysis in 1895 and the gene editing method CRISPR in 2012, was immediately known around the world. Other methodological discoveries are difficult to spot as they can be driven by multiple advances over centuries. The development of mathematical and statistical methods reflects such slow and quiet *methodological revolutions* that did not make international headlines. But they have fundamentally transformed how we do science and how we understand the world.

Statistics and mathematics are arguably the two most widely used methods across science, with statistics involving the collection, analysis and interpretation of data. Assessing over 750 major scientific discoveries, including all Nobel-Prize discoveries, illustrates that, since 1900, 99% of discoveries have used mathematics and 77% used statistics.[(64)] In physics, the field's two central theories are quantum theory (which incorporates probabilities and exhibits indeterministic behaviour) and relativity theory (which is described by mathematical formulas and is deterministic). In economics, the field is divided into applied economics (using statistical analysis) and theoretical economics (using algebraic equations). Most scientific theories are formalised using algebra and calculus, such as the equation $E = mc^2$. In many fields of science, inferential statistics (the analysis of data to make inferences about phenomena) has however become synonymous with the scientific method. Inferential statistics has revolutionised empirical sciences like biomedicine, psychology and social science by enabling us to generate causal evidence. Not only in science but also in our everyday life we often use statistical reasoning: we ask what is the likelihood of a medication we take to work, or our planet's climate to increase by 3 °C within a century or getting lung cancer from smoking. Statistics and mathematics are at the foundation of most of science. And like computers, it is difficult to imagine what contemporary science would look like without them.

We humans, with our limited cognitive resources and processing capacity (Chapter 4), have developed statistical methods that we apply together with computers to analyse hundreds and even millions of observations in a single

Science of Science. Alexander Krauss, Oxford University Press. © Alexander Krauss (2024).
DOI: 10.1093/9780198937401.003.0014

dataset. Without these methodological tools, we would not be able to study the world in such ways, and the advent of computers was key in quantifying and digitalising nearly all scientific fields (Chapter 12). Modern statistics transformed empirical science by allowing us to study the world with vast amounts of data in more complex ways and conduct and analyse larger-scale experiments. We apply statistical methods to study basically any phenomenon in science, from cells and viruses in populations, to planets, economic markets and science itself. In general, experimental science involves collecting statistical data about a given phenomenon from a sample, generating results and often estimating causal relationships, and then developing an explanation or theory for the evidence. Science commonly aims to establish cause–effect relationships in the world. And our understanding of causation is inherently statistical in experimental science: the more data we collect to observe a given phenomenon in different contexts and time periods, the more reliable our findings become.

Modern statistics developed in stages largely around the early 20th century and culminated in the landmark book *Statistical Methods for Research Workers* published by the British statistician and biologist Ronald Fisher in 1925.(193) It marked the first full-length book on statistical methods and was critical in establishing and spreading modern statistics. In this seminal book, Fisher developed some of today's most important and widely used statistical methods by scientists across all fields. He created the analysis of variance (ANOVA) technique that enables us to analyse the differences among subgroups or subexperiments within an experiment. It namely enables assessing the statistical differences between the means of the groups of a study sample—for instance of individuals or agricultural plots receiving different treatments. Fisher also popularised the use of the p-value, or probability value, in statistics, and he proposed using a 1 in 20 probability of a result arising by chance as a threshold for testing the statistical significance of experimental results (null-hypothesis testing).(193) His proposed threshold has become a universal constant. Today, most academic publications report the statistical significance of their results using his threshold, although debates are still ongoing about whether 0.05 is the best threshold (Figure 13.1). In light of the essential role of statistically rigorous research designs in shaping later study outcomes, Fisher stated that 'To consult the statistician after an experiment is finished is often merely to ask him to conduct a post mortem examination.' Fisher also pioneered the use of randomisation in agricultural experiments to reduce bias by randomly (rather than arbitrarily) selecting a sample from the general population.

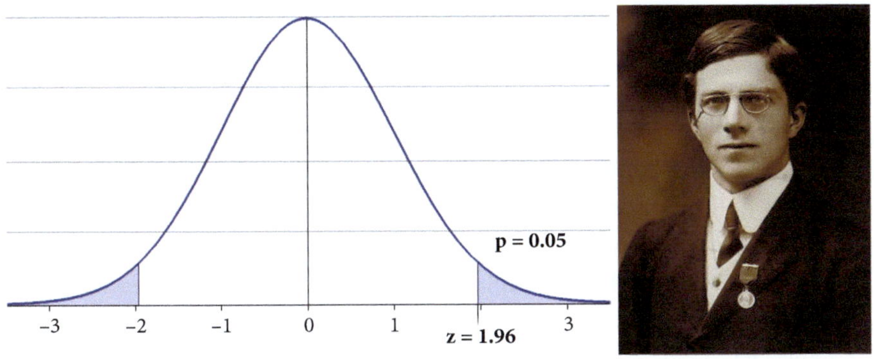

Figure 13.1 Most experimental results in science are tested with a threshold of statistical significance of 0.05 (1 in 20 odds of a result arising by chance), as proposed by Ronald Fisher.
Source: photograph from Wikimedia Commons, 1913.

A central limitation of statistical and mathematical methods is however that many aspects of the world cannot be easily studied using them since they cannot be captured well numerically. These range from ecosystems and many long-term health issues to psychological states and political institutions. Quantitative measurement is generally viewed as a central feature of scientific rigour and replicability. And robustness is a feature of measurement tools like statistics that enable us to test and reproduce results. Scientists generally direct their attention to those phenomena in the world that are most quantifiable—and thus measurable and amenable to statistical analysis. The level of quantifiability can range widely. On the one hand, there are highly stable physical phenomena (such as measuring gravity and the volume of certain objects) and phenomena that only exist as numerical constructs (such as age and income). On the other, there are not directly quantifiable phenomena (such as complex ecological systems and biological evolution) and historical events (such as the given year of the extinction of dinosaurs or the political causes of the fall of the Berlin Wall). Where a phenomenon falls on this spectrum determines whether we can more easily measure it (and thus whether it often appears more scientific) than other phenomena.

We make our observations of phenomena quantifiable and amenable to statistical analysis to be able to study them. We must fit observations of the world into variables (using for example binary values like 0 or 1) to be able to calculate statistical relationships between phenomena. Such binary variables we create capture an aspect of the world in quantitative terms. Creating such variables influences the phenomena we study by altering how we measure and observe the world. For phenomena in the human sciences are not naturally binary. No traits or behaviour fall squarely into a black or white description of the world. A common measure in medical studies is for example mortality, with which the effectiveness of a treatment is assessed by the duration that a study participant survives while taking the treatment. Yet such a binary measure does not capture participants' quality of life or level of pain, including the extent to which they suffer adverse events or are hospitalised before passing away.[(42)] This is a typical example of how scientists' value judgements enter into scientific evidence depending on how we define a variable used in a study (Chapter 6).

There are a number of problems that commonly face statistical studies and affect the quality of our evidence. Common problems in producing reliable and replicable scientific results are small sample size and low statistical power,[(194)] p-hacking and selective reporting, small effect sizes and HARKing.[(29,40,162,195,196,197)] Small sample size and low statistical power can affect studies negatively by increasing the chance of false positive results. P-hacking occurs when researchers for example collect additional data after assessing the statistical significance of results or exclude some outliers to improve the statistical significance of results. It arises given pressure to report only statistically significant or positive results, since journals are less likely to publish studies with statistically insignificant or negative results (publication bias). HARKing arises when researchers present unexpected results as if they were the initial intended hypothesis of the study, or present exploratory research as if it was confirmatory research (i.e. testing a pre-established hypothesis) (Chapter 5). Such research practices often lead to statistical biases in studies. This includes sampling bias (when individuals in a study sample are not representative of the broader population) and measurement bias (when data in a study are inaccurately measured or classified).[(29,42,162,194,195)]

While such biases and constraints face some studies, a broader set of complex biases and constraints generally face to some extent all statistical studies across the biomedical, behavioural and social sciences. A degree of variation between outcomes of different studies is expected given different conditions between people, contexts and time periods in which statistical studies are conducted. This is because researchers infer causal evidence from estimated statistical results that are the outcome of multiple complex processes. In the human sciences, researchers require selecting a sample, randomising, blinding and controlling the sample of participants, and carrying out treatments. Studies involve many actors, who include study designers, all participants, data collectors, implementing practitioners and study statisticians. These actors make multiple decisions at different steps when designing and implementing studies, and when analysing and extrapolating their results within a particular context and time period. Some difficulties in producing replicable statistical studies thus generally arise across the human sciences, since they are conducted by collecting data in a particular context (such as a clinic, laboratory or field) with particular people and conditions, and the phenomena being studied often change across contexts and can evolve over time. Replication studies are thus generally not able to obtain the same results (with the same effect sizes and significance levels) as in the original studies.[(198)]

The replication crisis across the human sciences—a widespread inability to replicate existing studies using new data—is thus partly driven by differences in contextual factors across studies, together with problems in using and reporting statistical results.[(198)] This constrains 1 to 1 replication from one study context to another.[(42,162,196)] All phenomena, when measured statistically, face some degree of variability, and researchers generally report in studies the degree of precision and error bars (that estimate the degree of error) for their data.

An important related dimension of science is the geography or context of science that deals with the generalisability of evidence beyond the context in which we collect data. It deals with the external validity of our evidence—the conditions under which, and extent to which, data we gather in one context apply to another context. We live in a limited pocket of the universe (the planet Earth) and much of the universe is inaccessible to us and our best telescopes and spectrographs. We also live at one point in time, within our current century, and cannot directly collect data or run experiments on the past. For us to do science and create knowledge depends on our geographic and temporal context in which we can gather data. We conduct our studies in a medical clinic, a geological site, an ecosystem and our universe—that is, within a particular geographic location and time period.

Our most reliable knowledge is commonly viewed as knowledge that transcends context, that is universally valid. Examples include that living organisms require energy in the form of food and that electrons are negatively charged. For most phenomena we are however not generally able to collect data and run experiments universally but rather only locally, within particular laboratories or fields, especially across the human sciences.[(199,200)] The data generated in an experimental context are the basis for the scope of results and causal claims we can derive from the experiment. Our evidence becomes more reliable the more data we collect across studies conducted in different contexts and at different times. Our evidence also becomes more reliable the more diverse methods we apply that produce comparable results supporting existing evidence.

These issues are related to the external validity of evidence that we illustrate here with the method of RCTs. RCTs are commonly viewed as the leading method across the biomedical, behavioural and social sciences for assessing how effective a medical treatment or policy intervention is. RCTs estimate the causal effect of an intervention and we often face some limitations in using those estimated causal effects outside the study context. If for example a medical treatment for diabetes reduces blood sugar levels by 5% on average among treated participants in a trial study, it can be difficult to attain the same effects among the general population after the trial. This is because of demographic and clinical differences between study participants and later the broader population for which the intervention is actually intended. For study participants are in a controlled trial context while the general population does not undergo experimental controls.(201) Limitations also arise due to differences in the extent to which institutional, economic, educational and behavioural conditions are met. In no two contexts are all conditions met to the same extent for a treatment to produce the same causal effect. When we use results derived from a trial population in a different population, we always face some constraints given variation between populations. A trial's estimated causal results are relative to particular factors under particular conditions within a particular sample at a particular time.(42)

How we explain evidence thus depends on how and where we collect the data to do so. How we express our evidence and knowledge is influenced by the context in which we collect data as well as by our human mind and senses, methods and practices, categories and definitions (Chapters 2–4, 10–15). We do not have a perspective from nowhere through which to gather data that transcends time and space, and that is independent of our methods. Yet the more data we collect in different contexts and time periods, and the more methods we use, the more reliable and robust our knowledge becomes.

In general, methods are typically passed along to the next generation who were not exposed to the debates that had to first take place before those methods became accepted. Methods however face assumptions, limitations and potential biases that are not generally outlined in studies that use them (Chapter 10). At many universities, graduate students in fields like biology and medicine are generally only required to take at most one course in research methods like statistics. But later they require using those methods to publish their results and they may not be aware of the range of assumptions, limitations and potential biases. This can constrain research quality.

Let us turn now to steps we can take to improve research and its potential impact. Because science relies heavily on statistical methods, the quality and replicability of our evidence also depends on how we design our statistical studies, refine our statistical methods and report our statistical results. Improving research practice by conducting studies in multiple contexts, pre-registering studies, better blinding and randomising participants in studies, and updating reporting standards for studies are among the important ways we can reduce bias. Pre-registering study designs can for instance be an effective way to blind participants and reduce bias, as the data and outcomes do not yet exist and cannot be later altered. Pre-registration, while common in clinical medicine, is not always feasible. And importantly, most scientific research will remain exploratory, not confirmatory, and so cannot be pre-registered.(195) We can also make experimental designs more robust by building greater rigour and replicability into the research design. In the human sciences we can achieve this by integrating comparability of groups, treatments and contexts within a single study—that is, building replicability into our study designs.(198) Research needs to be not only statistically robust and reliable but also, with most research publicly funded, useful for society or policy (Chapter 7).

Taken together, such statistical issues have contributed to a replication crisis in science. Methodological causes of the crisis include small samples and effect sizes in studies, p-hacking and statistical standards including the statistical significance level that researchers apply in their studies.(162,195) Structural causes contribute to the crisis that are related to scientific institutions and journals, including the current reporting guidelines for studies, and the priorities of funding agencies and reward systems (Chapter 7).(162) Psychological causes of the crisis arise due to confirmation bias, heuristics-based reasoning and our bounded rationality (Chapter 5).(109) Just as the replication crisis within science is driven by a complex of methodological, structural and psychological causes,(202) science in general is driven by such a broad range of factors, outlined throughout the book. But each science of science subfield has focused on one dimension. The scientometric community and history of science community for example have largely not taken into account the range of methodological and psychological factors shaping science. The best way to understand and address the diverse causes of the replication crisis is also the best way to understand and advance science: by adopting an integrated approach.

In sum, methods like statistics and mathematics are fundamental to how we conduct science. Yet we face constraints when using such methods and making phenomena in the world amenable to them. Challenges arise in replicating statistical studies as we face difficulties in designing, implementing and analysing studies in different contexts, especially in the human sciences. The geographic location and time period in which we gather data also shapes the parameters, content and scope of our evidence and knowledge, particularly when dealing with human beings. We need to thus conduct studies in multiple contexts to make our evidence more reliable. Much ongoing debate on reducing biases and enhancing quality and replicability of our scientific evidence revolves around improving statistical methods and reporting of studies. Our statistical and mathematical methods thus help shape the foundations, present limits and advancement of science, and account for central floors in our tower of science. Since its birth, statistics has been the practical solution across science to the problem of identifying causal relationships in the world thanks to statisticians like Fisher and the practical solution to the problem of induction that philosophers like Hume aimed to address, which we turn to next.

14
Philosophy of Science

What science is and its foundation have been explored by philosophers for centuries, including Francis Bacon, David Hume and Karl Popper, and what knowledge is and its foundation for over two and a half millennia, including Plato, Aristotle and Ludwig Wittgenstein. Philosophers have addressed central questions of science of science longer than researchers in any other subfield. Major debates that have dominated philosophy of science include paradigm shifts, justification, induction, demarcation and realism, as evidenced in leading philosophy of science textbooks.[18,19,203,204,205] Paradigm shifts refer to fundamental changes in the theories of a scientific field.[1] Justification deals with principles such as falsification and verification to justify our theories of the world (and in statistics it refers to falsifying 'by chance' hypotheses).[15,33,107] Induction addresses the question of whether observations we make can or cannot justify generalising about the observations in other contexts or in the future, which is called the problem of induction (and in statistics it refers to the problem of external validity of results).[206,207] Demarcation involves defining criteria for what is and is not science.[15,16,130,208] Realism concerns whether scientific theories provide a reliable approximation and true description of reality, for observable and not directly observable phenomena.[209,210]

In general, what we can know, as human beings with cognitive limitations, has arguably been the central question of epistemology since Hume and Kant in the 1700s.[206,211] These philosophers stressed the foundational role that our mind plays in how we express our ideas about the world. Yet there is still no consensus among philosophers on which explanations of knowledge and science are most accurate. The reason is because philosophers have theorised about what knowledge and science are mainly using conceptual analysis, individual case studies or abstract normative methods rather than more systematic scientific methods that allow for generalising about knowledge and science.[15,16,107,206,207,210,211,212,213]

In the 4th century BCE, Plato famously described knowledge as 'justified true belief.'[212] Most philosophers (and nearly all humans) view observation as the main source of our knowledge of the world. It is the common

Science of Science. Alexander Krauss, Oxford University Press. © Alexander Krauss (2024).
DOI: 10.1093/9780198937401.003.0015

sense view throughout human history going back to early *Homo sapiens* (Chapters 3 and 4), stressed by the ancient Chinese,[157] outlined in detail by Al-Haytham in 1021, and reiterated again by Robert Grosseteste, Roger Bacon and Thomas Aquinas in the 1200s, and later Francis Bacon in his book *Novum Organum* in 1620[214] and then by positivist philosophers like David Hume and John Stuart Mill, among many others. The empiricist view is thus that our sensory experiences in the world are the basis of our ideas and what we can know. That is, nothing goes through our intellect that did not first go through our senses (Chapters 2–4). Francis Bacon stressed the need for studying and intervening in reality by conducting experiments and empirically testing reality (or, as he said, by 'twisting the lion's tail' and seeing what happens),[214] although this did not become the standard approach in philosophy, as it did in science. In the 1920s, the philosophers of the Vienna Circle again reemphasised the importance of observation but went beyond the positivist focus on the strictly observable world. They also stressed the need for using methods of verification, mathematics and logic when creating and justifying knowledge (Chapter 13).[18,107]

Explaining how we create knowledge and how science operates are in philosophy at times reduced for simplicity to two theories of scientific methodology: induction and falsification. Bacon's theory of induction and scientific methods is commonly viewed in philosophy as the conceptual origin of modern science, namely using the methods of observation, experimentation and deriving conclusions.[214] These are the same methodological abilities of the mind that our species have always used and will continue to always use to acquire knowledge of the world, and we have done so in increasingly systematic ways over time (Chapters 2–4). In general, inductive reasoning is when we for example repeatedly observe the sun rising and then infer that it will rise tomorrow or even that it will always rise every morning. It allows us to go beyond our current set of observations and draw conclusions about the future and is directly linked to how we make statistical inferences across science (Chapter 13).[206,207,214]

For falsificationists, in contrast, scientists need to do science by constructing hypotheses and theories, testing them and attempting to falsify them.[14,15,215] Popper argues that the defining trait of scientific investigation, evaluation and justification of theories is the principle of falsification—which is the most influential account in the philosophical literature and most well known outside of philosophy.[215] As Popper stated, 'The growth of knowledge depends entirely on disagreement.' Scientists have not however adopted falsification as a guiding philosophical principle for evaluating and justifying theories. While falsificationists make theories the foundation of their inquiry, inductionists focus on the process of observing and experimenting while trying to reduce underlying assumptions and theory.[18,19] For both inductionists and falsificationists however, theories like Freud's psychoanalytic theory are not scientific as we cannot easily empirically test or falsify them.

These two central theories of scientific methodology in philosophy (falsification and induction) have a long tradition in statistics. For statisticians there is not much new to Popper's principle of falsification. The notion of falsifying hypotheses was first developed in statistics and has a long practice rooted in Fisher's null-hypothesis testing and falsifying 'by chance' hypotheses (i.e. not proving but disproving hypotheses). This is one of the most widely used statistical techniques across all scientific fields today, and was outlined in Fisher's seminal book in 1925 (Chapter 13).[193] For statisticians, what they call the problem of external validity is, for philosophers, also known as the problem of induction. For both, it refers to the constraints we face in generalising from observations we make to observations in other contexts—for example, extrapolating results from a particular experimental sample of data to a broader or future population. Statistical methods have been developed and are widely applied across scientific fields to address both falsification and induction—such as null-hypothesis testing, random and stratified sampling, and replication of studies (Chapter 13). As philosophers of science—like other researchers in science of science subfields—work largely in isolation, they have not yet resolved these two major debates by adopting the practical solutions long developed and used in other fields, although thousands of articles in statistics exist on how to address these two issues. Another related central debate in the philosophy of science is on demarcation (identifying criteria for what is and is not science). We will later discuss the topic and provide a new criterion for defining the boundaries of science (Chapter 19). Philosophy of science moreover covers domains such as philosophy of physics, biology and economics that provide insights into the complexities of these fields' concepts, definitions and assumptions.

We can also better understand science by comprehending the metaphysical aspects underlying science—that is, digging under the surface of science to identify its ontological aspects. Metaphysics of science is a branch of philosophy of science that can be classified into two broad categories: on the one hand, it studies causal and mechanistic explanations, measurement of causes and the nature of scientific regularities/laws; on the other hand, it studies the metaphysical assumptions of science inherent in research, namely in methods, definitions, concepts and theories.[216,217,218,219]

We observe causal relationships for example in the different degrees of stability found in phenomena in the world—protons are positively charged; a balanced nutrition improves health; and so on. For each proton or person's health we assess, we do not need to start from scratch by conducting experiments to identify whether protons are positively charged or a balanced nutrition improves health. Identifying regularities and causal relationships enables us to explain and predict them.[216,217,218] Across scientific fields, we measure these using different methods. Each method generally provides us with a different perspective to better understand the given phenomenon. In studying consumer behaviour or depression for example, psychologists and applied economists generally apply statistical and experimental methods, sociologists conduct qualitative studies and interviews, and theoretical economists use deterministic mathematical equations. Each group studies the same phenomenon but captures it in different ways (Chapter 13). How we measure cause–effect relationships in the world is embedded in and are properties of the particular methods we apply.

When we describe regularities in the world there is also often a trade-off between simplicity and strength of an explanation: the greater the simplicity used to describe phenomena, the greater the loss in power to explain them.[220] The best explanation, theory or 'law' would ideally account for strength and simplicity. Newtonian physics for example has in general less strength but is simpler than quantum physics. For the philosopher Nancy Cartwright, the fundamental principles and theories in physics provide an idealised representation between abstract theory and empirical reality. Also, scientific theories and laws are not universally applicable, as Cartwright argues, but are true only *ceteris paribus* (all other things being equal) within a given context.[220]

In general, different scientific communities develop their own ontologies (such as definitions and concepts), systems of measurement (such as descriptions of quantity and mathematical representations), causal classifications (such as deterministic, probabilistic and ceteris paribus causal relationships), levels of evaluation (such as the property or system, or process or outcome) and thus epistemologies (explanations, theories and worldviews). These different features are embedded in each scientific method we develop and use. Our methods and instruments (experimental, statistical, chromatographic) shape the metaphysical aspects of science through these different domains. In this way, our tools of science standardise and automate features of conducting science (Chapters 10, 12 and 13).

We illustrate this point here with RCTs. When we use the RCT method, we automatically focus on studying an isolated single causal effect (the targeted intervention) instead of multiple causal effects within the larger context, which we cannot study using RCTs. We focus on quantitative instead of qualitative data and on the outcome instead of the process of the phenomenon. We also focus on the average causal effect instead of the distribution of effects, on simple instead of complex interventions and on an isolated instead of holistic explanation of phenomena.(42) It is the features of a method and research design that thus shape our ontological views, measurement practices, causal understanding and evaluation criteria. In turn, these shape our epistemology of the world. The RCT method embodies different features within these different domains, compared to other methods such as observational studies or individual case studies used to study similar questions. While the RCT method has transformed how we understand interventions targeting disease, human behaviour and public policy, it neglects other aspects of the same phenomena that are better captured by those other methods. Overall, how we understand phenomena cannot be viewed separately from the way we methodologically measure and represent them. Our scientific methods and instruments are our lenses through which we perceive the world. An important implication for science is that we need to apply multiple methods so we can better understand different aspects of the same phenomenon using different evidence and improve our aggregate understanding of the phenomenon—the approach adopted in this book.

Turning now to the metaphysical assumptions of science, scientists rarely outline the assumptions implicit in their research and the methods they adopt. These can be classified here into seven main traditional assumptions underlying science that guide how most of science needs to be done and are inherent in our common conceptions and definitions of science: *mathematical*, *material*, *reductionist*, *universal*, *causal*, *unidisciplinary*, and *one-scientific-method*. These traditional assumptions have been important in bringing about medical, technological and scientific achievements. They guide scientific research and specify what counts as scientific. These assumptions did not however always exist. They have been adopted in a piecemeal fashion increasingly since the 17th century and are observed for example in the works of Kepler, (221) Galileo,(222) Newton(223) and Einstein.(224) Scientists are not always aware that they are assumptions. They often view them as established scientific facts needed to do science, as observed in scientific publications. Physicists commonly adopt most, if not all, of them—as for example captured in $E = mc^2$ associated with Einstein's theory of special relativity or in $F = ma$

associated with Newton's law of motion.[223,224] These assumptions provide methodological rules for doing science, but they all face exceptions:

- *The mathematical assumption*: Scientific method is the 'mathematical and experimental technique employed in the sciences [and] used in the construction and testing of a scientific hypothesis,' as defined by Encyclopaedia Britannica.[225] Nature thereby speaks the language of mathematics.[221,222,223] However, many phenomena cannot be measured well mathematically—from evolutionary processes and historical events, to molecules, viruses, anatomy and botany.
- *The material assumption*: Science studies 'the nature and behaviour of the material and physical universe,' as defined by Collins English Dictionary.[226]cf.[65,227,228] However, many phenomena are not entirely material or physical—from consciousness and space and time, to scientific reasoning, institutions and social norms.
- *The reductionist assumption*: Science aims to reduce the complexity of the world in the simplest way.[229,230,231,232] However, many phenomena cannot be reduced for example to mathematical equations and statistical models that can neglect the complexity of such phenomena—from ecological, biological and social systems, to economic reforms.
- *The universal assumption*: Science studies the world with the aim of establishing universal facts and formulating 'laws to describe these facts in general terms,' as defined by Collins English Dictionary.[226]cf.[228,233] However, many phenomena do not have a universal structure but are context-specific—from natural catastrophes and evolutionary processes in biological organisms, to financial markets.
- *The causal assumption*: Science studies cause–effect relationships and all phenomena underlie them.[233,234] However, many phenomena do not exhibit a discernible cause and effect—from gravity and human evolution, to weather patterns and psychological states.
- *The unidisciplinary assumption*: Science is conducted with each academic field studying a different domain (as discussed in the Introduction).[235,236] However, we best understand most complex phenomena from across disciplinary perspectives that provide different evidence by applying different methods—from climate change studied by physicists, geographers, political scientists and others, to human behaviour studied by evolutionary biologists, cognitive scientists, economists and others.

- *The one-scientific-method assumption*: Science requires applying the scientific method based on 'the collection of data through observation and experiment, and the formulation and testing of hypotheses,' as defined by Merriam-Webster Dictionary.[237)cf.(214,226] However, some phenomena cannot be studied using these methodological approaches—from not being able to conduct experiments on fossils, planets or the development of cities, to not being able to test hypotheses when surveying rock formations or describing historical events.[238]

Given the great success of physics in explaining vast parts of the world—in terms of relativity theory and quantum theory—these assumptions have become an ideal for researchers. Researchers in other fields generally adopt several or more of them. The lens through which we study the world—the lens of mathematics, reductionism, unidisciplinarity etc.—portrays atoms, the universe, economic actors and markets as mathematical, mechanical, unidisciplinarily etc. These foundational assumptions help frame the questions we ask, the methods we use, the way we interpret data, the theories and models we develop. These common assumptions of science, when viewed as how science should be done, can however constrain science. With an awareness of these assumptions, we can open up science and adopt alternative methodological approaches (such as non-mathematical and cross-disciplinary) to address many questions and challenges we face. Despite the various philosophical insights into the nature of science, philosophy of science faces a number of shortcomings.

Overall, philosophy of science has, to date, not yet provided a comprehensive understanding of its subject matter—science—for four reasons. Firstly, most literature in philosophy of science has focused on studying theories (one output of science) and has done so focusing on the field's major debates of paradigm shifts (in theories), justification (of theories), induction (for developing theories), demarcation (for testing whether theories are scientific) and realism (whether theories provide a reliable approximation of reality), as observed in leading textbooks in the field.[18,19,203,204,205] That philosophers of science focus on one output of science—theories—is understandable because their knowledge of science is centred mainly around this scientific output and they do not commonly work in labs and conduct experiments using scientific methods (the process of science). The field has not given sufficient attention to the fact that its main subject of study, science, is developed and advanced by creating new scientific instruments and methods, such as novel X-ray diffraction methods, computational methods, electrophoresis, statistical methods, particle accelerators and chromatography. And these particular methods and instruments of science have not received much, if any, attention in terms of studying their assumptions, biases and limits and how to push those limits to drive new discoveries (Chapters 10, 16–20). Related research has instead focused on general scientific methodology, in abstract terms. Leading philosophers of science who have led the major debates in the field outlined above have thus not focused on the major methods and instruments of science as the main mechanism driving new scientific advances, theories, discoveries and fields.[(1,15,33,206,210)cf.(18,19,203,204,205)]

Secondly, most leading publications in philosophy of science have studied what science is using theoretical, conceptual and normative approaches, rather than also using systematic (scientific) methods from across scientific fields—experimental, statistical, computational etc.[(1,15,33,206,210)cf.(18,19,203,204,205)] Thirdly, most philosophers have used their mind as the main source of acquiring knowledge. They have largely not taken into account the powerful role of systematically gathering empirical data and carrying out representative studies to make robust and generalisable inferences—as used in all fields of science.[(1,15,33,206,210)cf.(18,19,203,204,205)] Fourthly, what most publications in philosophy of science have studied and the proposed explanations are commonly an internal response to philosophical questions. They are not commonly a response to problems facing contemporary science, including those related to methods, instruments, biases and institutions that contributes to existing scientific literature.[(ibid.)]

The Nobel laureate in physics Richard Feynman claimed that 'philosophy of science is as useful to scientists as ornithology is to birds.' The philosopher Hasok Chang also argued that 'most scientists today would regard most discussions currently taking place in professional philosophy as utterly irrelevant to science.'[(239)] Such claims appear to also include the philosophy of science 'in practice' that generally does not build on the vast existing scientific literature. This can be seen by looking at a sample of philosophy of science publications—via Scopus or Google Scholar—and exploring who they cite and who cites them, namely other philosophers. Yet it can seem surprising for an outsider that philosophers of science study science, without aiming to influence or improve science and thus to also publish their studies in science journals. Can internal dialogue with other philosophers be the purpose of philosophy of science rather than engaging with and informing scientists and helping improve science?

The old epistemic questions about what knowledge, science and their foundations are and how we develop them, which philosophers have studied for millennia, can be more rigorously answered scientifically using empirical evidence (Chapters 2–15). And only then can we develop a coherent concept and understanding of knowledge and science that is consistent with existing empirical evidence about knowledge and science across fields and grounded in scientific practice (Chapters 16 and 17). This requires integrating a range of methods and evidence to be able to arrive at conclusions that are coherent with (do not contradict) what is already known across scientific fields. It is for these reasons why the field of philosophy of science, which raises some of the best questions about the foundations and limits of science, has to date largely not been able to address those questions. This traditional philosophical project, by adopting a unidisciplinary approach to studying science, has not sufficiently taken the empirical world, scientific methods and actual scientific practice into account—at least not in a way comparable to the rigorous methods applied across science. Philosophy of science is another case in point illustrating why we can only understand science comprehensively by taking an integrated approach. The integrated account adopted here can provide a more empirically founded and scientific understanding to these foundational questions about science (Chapters 2–20; Figure 1.2).

In studying the nature of science, the key difference between the old field of *philosophy of science* and the new field of *science of science* is that the former relies mainly on theoretical and conceptual analysis, while the latter relies on empirical methods and evidence. Providing scientific responses using scientific methods to old philosophical questions is how many scientific fields, throughout history, emerged out of natural philosophy. This is also the case for the field of science of science outlined here. The general trend has however changed. Throughout history, science has often begun where philosophical questions ended. Philosophical questions now often begin where science ends. While scientists push forward and continually seek to expand our understanding of the world, philosophers of science commonly seek to demonstrate problems and limitations inherent to science and our limited understanding of the world. It is often a difference between academic optimists and pessimists.

Finally, often 'a scientist formulates problems in a way which requires for their solution just those techniques in which he himself is especially skilled.'[(240)] Researchers may often express the questions they address in such a way that the answers to those questions can be best provided with the methods that they already have at their disposal, as opposed to methods in other fields that they may not be familiar with. If scientometricians' main method is large-scale data analysis, if philosophers of science' main method is conceptual and theoretical analysis, and so on, then questions look like those that only large-scale data or only conceptual and theoretical clarity can address. This provides another perspective to the disciplinary isolation among the science of science community to date—a community that collectively studies science (when viewed from outside) but is largely not aware that it is part of this much larger community across many fields aiming to address the same questions (when viewed from within each subfield). It also provides a further explanation for why the science of science community has not yet jointly adopted multiple methods and a truly interdisciplinary approach that stretches across all fields, as outlined here.

In sum, over millennia, philosophy has contributed to understanding the theoretical and conceptual foundations and boundaries of what we know. Philosophers of science have widely debated how to theoretically explain and justify knowledge and science, especially using methodologies like induction and falsification—which statisticians have long provided practical solutions to. Such philosophical debates often revolve around theoretical and normative principles of what knowledge and science should look like and how to demarcate them—that is, principles of what grounds the tower of science, what its borders are and what shape it should take. But such theoretical principles have not generally been directly used when doing science and applying methods in practice. Science also involves studying causal relationships, mechanisms and regularities, and adopting metaphysical assumptions. Understanding these provides insights into the foundations of science and can improve how we do science by expanding our methodological approaches and how we view and describe the world. And as the philosopher Wittgenstein stated, 'the limits of my language are the limits of my world.'

15
Linguistics of Science

We turn now to the role of language in science and understanding science. Without a system of language we would not be able to reason complexly, express our knowledge and do science (or even read books on science like this one). It enables us to describe and explain to others what we observe, how we solve problems and the knowledge we acquire about the world.[21,90,92] With language we can quickly obtain and pass along methods and bodies of knowledge. How we use language determines how scientific studies are expressed and disseminated and how accessible they are to researchers in the same and other fields—or whether research can only be accessed by few.[21,90,92,52,87]

Written and especially digital documentation allows us to more efficiently share and cumulatively build on vast bodies of knowledge and methods across generations (Chapter 6).[21] It frees up our limited working memory by making it much easier to process information such as multiple ideas simultaneously and quickly expand on them. It helps us address complex problems and questions more effectively. It also extends our long-term memory to systematically record what we observe and to store knowledge. A system of written language is a precondition for cumulative knowledge and creating scientific methods including a system of mathematics.[241] Only by using language can we express our methods of science, including statistical coefficients and algebraic equations. Our methods and knowledge cannot thus be independent of the language we have developed.

Technical language divides the scientific community in general, including the science of science community. A specialised language connects researchers in one subfield with a common language but often presents a barrier for researchers in other subfields to understanding the content of studies. Some fields are laden with technical jargon. Think of common terms used among economists of science, such as elasticity or hyperbolic discounting (which likely cause a blank face when outsiders read them). Elasticity refers to how one factor (such as demand for knowledge) responds to a change in another factor (such as supply of government funding). Hyperbolic discounting refers to delayed discounting or decreasing returns over time for choices we make (such

Science of Science. Alexander Krauss, Oxford University Press. © Alexander Krauss (2024).
DOI: 10.1093/9780198937401.003.0016

as related to financial rewards for making future breakthroughs). Think of common terms among philosophers of science, such as the metaphysics of science or logical positivism. For non-philosophers, it is difficult to imagine what they could mean. Metaphysics of science studies questions about fundamental concepts, categories and definitions used in science. Logical positivism refers to a philosophical movement to use observation together with logical methods to generate and justify knowledge. Think also of different systems of language used to describe evidence across science of science subfields, with results of different studies presented using for example statistical programming language, computational language, language of conceptual analysis, mathematical formalism and other methodological languages. Researchers speak one or several of these languages but not others. It will be key for researchers in science of science to begin simplifying language, leaving out or simply defining field-specific jargon in studies. For jargon constrains other researchers to understand studies and build on research, in order to advance the common objective of the science of science community of developing a coherent understanding of science.

At present the English language dominates science worldwide, including the leading scientific journals and institutions. This presents a challenge as most literature across science of science generally only studies articles published in English and so can provide an incomplete picture of science. Large-scale scientometric studies for instance commonly only capture articles in English. If we want to understand for example what drives major discoveries, including Nobel-Prize-winning discoveries, this can require using digital translation programmes to extract necessary data since those discoveries were published in over a dozen languages.

Language and writing systems—like scientific methods—are central thinking tools. The Western alphabet is viewed as structured analytically and a natural tool for categorising. It functions as a model for classification systems, and standard measures and weights. The Chinese writing system, on the other hand, is largely pictographic and non-reductionist. It functions as a model for viewing the world as continuous and holistic, and facilitates taking the broader set of factors influencing a phenomenon into account holistically (Chapter 9).[(57,158)] Using a particular alphabet can thus shape the way we think and view phenomena.

Another important area in which language shapes how we use and disseminate science is the communication of science.[(242)] The Nobel-Prize-winning physicist Ernest Rutherford stated, 'An alleged scientific discovery has no merit unless it can be explained to a barmaid.' Findings of studies on topics like climate change, nutrition, vaccinations and the coronavirus pandemic are especially relevant to all of us, the general population. How science is communicated to the public, policymakers and other scientists is important as it can affect their decisions and behaviour. Scientists need to ensure that a study's results cannot be easily misinterpreted, that results of political, social and ethical relevance are presented sensitively and that uncertainty, risks and values are communicated in a balanced way. Uncertainty is a feature of many scientific fields. How uncertainty is communicated in studies (such as on environmental issues and consumer behaviour) is crucial, given possible negative effects on our choices.[(243)] Language used in social and digital media, news outlets and popular science books can be especially susceptible to misuse, as they do not undergo rigorous peer review but can still have a broad impact on the public. [(242)]

In sum, our language, especially written language, helps lay part of the foundation of science and our understanding of science by enabling us to reason, acquire knowledge and use methods more complexly and cumulatively. It also helps shape the way we express and disseminate our ideas and view the world, while it can create barriers between fields laden in technical jargon. The linguistics of science thus contributes to understanding part of the foundation of the *tower of science* and the divisions we observe across the science of science community. After discussing the 14 subfields of science and science, the next chapter synthesises the evidence from across the subfields to be able to provide a more comprehensive understanding of the origins, foundations and limits of science—and thus a holistic picture of our tower of science.

16

Science of Science: An Integrated and Methods-Driven Understanding of Science

We have developed science by using our cognitive and sensory abilities that have evolved within our environmental niche of the world and they face constraints, and we expand our vast knowledge of the world by developing new scientific methods and instruments designed to reduce our constraints. This universal and adaptive methodological toolbox of ours is at the centre of science and enables us to do and advance science in new ways. Other factors also influence science as we are social beings embedded in our scientific community and its practices, socialised into a system of language and mathematics used to express our knowledge, born into a historical context with particular world views, abiding by scientific norms, principles and assumptions, motivated by biological traits, subject to psychological biases, aided by computer technology and influenced by recognition and ambition, funding and societal objectives, public and economic institutions. Ultimately, science is a dynamic system of human activities aimed at better understanding the world.

What drives this system are complex interactions between our evolved mind and the methods we develop using our mind, on the one hand, and the world and social institutions, on the other. Science is the outcome of arguably the most cognitive and social activity that our species has undertaken. It is an activity in which we interact with our natural and social environment to develop sophisticated methods and instruments and bodies of knowledge that we cumulatively build on over centuries. What we call science is thus an organised effort of accumulating knowledge by working together to develop and apply increasingly complex scientific tools. These elements come together to make our ideas, discoveries and scientific fields possible. The elements lay the foundation, structure and form of our tower of science.

A more integrated and coherent understanding of the foundations and limits of science is thus inseparable from methods and evidence in cognitive science and methodology, biology and archaeology, computer science and statistics, anthropology and psychology, sociology and philosophy, economics and scientometrics, and history and linguistics. Scientists are members of a

Science of Science. Alexander Krauss, Oxford University Press. © Alexander Krauss (2024).
DOI: 10.1093/9780198937401.003.0017

scientific community shaped by its methodology, its history, its sociology and its philosophy. Artificial borders between disciplines have traditionally characterised science and the study of science. What were once thought of as independent disciplines—with tens of thousands of existing publications studying science from their own disciplinary perspective and methodological approach (Figure 1.1)—have been linked together here to account for science holistically (Figure 1.2). What enables and constrains science cannot be explained from a single disciplinary perspective alone. But almost all articles studying science do precisely that. As one indication, less than 3% of total publications across all subfields use the common terms inter-, cross- or multi-disciplinary (Chapter 1). To address foundational questions about science, some generally view the appropriate unit of analysis to be the individual (psychologists and cognitive scientists). Others generally think it is the group (anthropologists, sociologists, economists and linguists). Some generally study at the level of the species (biologists and evolutionary cognitive scientists). Others generally investigate the past (historians, archaeologists and some anthropologists). Some adopt a meta-level or methodological perspective (methodologists, scientometricians, computer scientists, statisticians and philosophers).

Yet disciplinary isolation has given rise to simplified and at times polarising views. Leading scientometricians and network scientists, like Fortunato, Wang and Barabási, have focused on and stressed the key role of publications and citations.[4,5] Leading historian of science, Kuhn, the changes in scientific paradigms.[1,32] Leading philosopher of science, Popper, the evaluation of scientific theories.[14,15,31] Leading sociologists of science, Latour, Woolgar and Bourdieu, the social practices of scientists,[10,11] and so on. This common approach to studying science has led leading researchers to not yet address the central questions of how important the particular 'key' factor they study is and how it relates to other 'key' influencing factors identified by other researchers across different fields that they do not study. It has led at times to overinterpreting the role of the factor they study compared to other factors, especially the foundational role of our scientific methods and instruments and our mind in enabling and constraining science—which leading researchers do not focus on (Chapters 2–15). Different researchers working at a different level and on a different aspect of science of science are driven by the common aim of understanding and improving science. But they generally do so in methodological and disciplinary silos, which has led to an incomplete understanding of science.

The unified account of the field of science of science presented here, by bringing together the approaches and features across 14 disciplines, outlines the evidence on science that is coherent across the natural, behavioural and social sciences. Taking such a holistic approach represents the most comprehensive understanding we have of science for the following reason: *the range of disciplinary approaches apply different methods and focus on different features of science, and there is coherence across the independent strands of evidence, in particular in the role of methodological features in shaping science* (Chapters 2–15).

To comprehensively understand for example how our mind enables us to do science, it is not enough to just study our mind from one perspective. We cannot just adopt a perspective from neuroscience using brain scans or a perspective from psychology running experiments with individuals solving

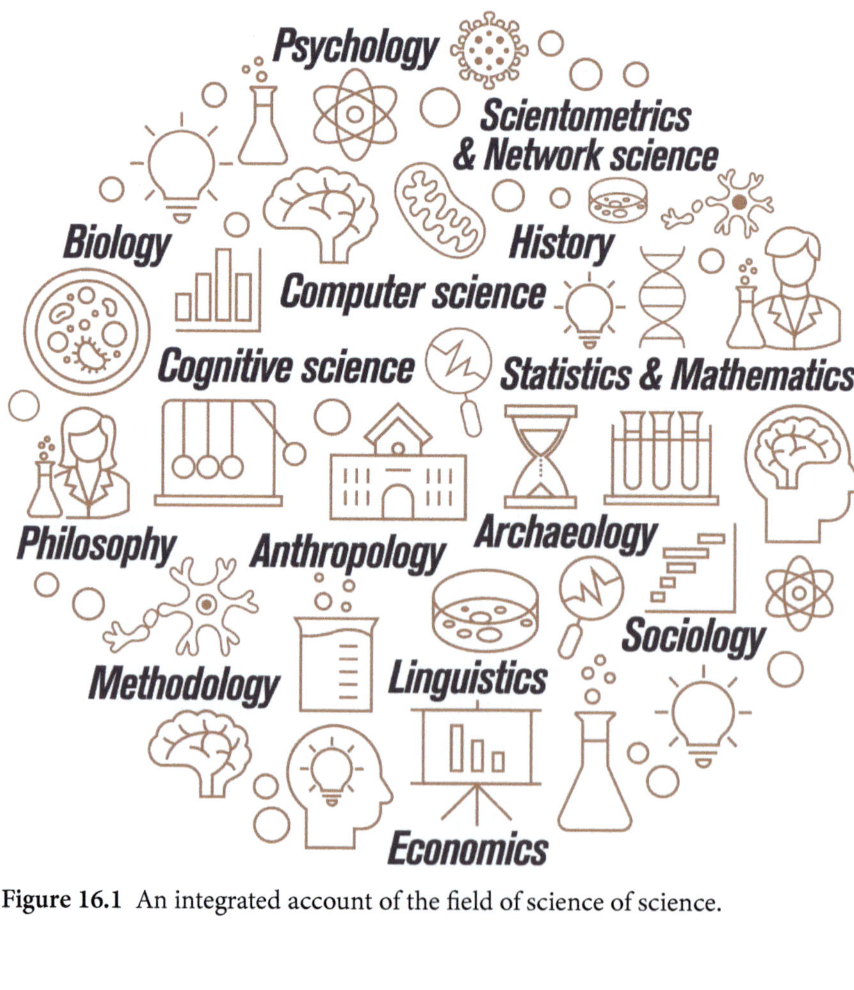

Figure 16.1 An integrated account of the field of science of science.

problems. Instead, we need to integrate such evidence together with evidence about our universal methodological abilities of the mind, which we share with other animals for observing and solving problems, which have evolved by adapting to our niche of the world and which we expand by developing new methodological innovations. We need to study how our mind is embedded in a cultural and historical context that shapes our different approaches to reasoning, as observed across scientific fields (Chapters 2–15).

As a simplified summary of the book, Figure 16.2 outlines the set of interconnected abilities and conditions that enable and constrain science and, in doing so, provides an integrated view of the field of science of science. It outlines what the field looks like when we study science from all perspectives. We and what we are methodologically capable of are at the centre of the foundations and present limits of science. Understanding ourselves and our cognitive and methodological constraints is the key to understanding science and the scientific methods we develop to address those constraints and expand the research frontier.

Explaining different aspects of the same phenomenon—science—at different levels enriches our overall understanding, as Figure 16.2 illustrates. We cannot comprehend the individual factors in isolation, because the parts (and the groups) interact with each other to account for the greater whole. Internal factors such as our cognitive abilities and senses shape which methods and instruments we develop and improve to make scientific advances. External factors such as science funding and the size of the scientific community influence which scientific methods and instruments we can develop, what research we fund and how many researchers work together. Meta-level factors such as our methodological limitations and assumptions shape the foundations of our research and how we can advance science, as we go about addressing our cognitive, methodological and instrumental constraints, and so on.

We cannot comprehensively understand the interconnected complexities of science without integrating the disparate evidence from across fields. The same applies to understanding scientific discoveries. Different aspects of the discovery process have been studied by different researchers—with some psychologists analysing imagination and analogies used in discoveries,(244) scientometricians and network scientists assessing age dynamics and collaborations among discoverers,(4) computer scientists running simulations of discoveries,(188) historians investigating theory change brought about by discovery(1) and philosophers examining how discoveries are explained and justified.(14,15,215) It seems a natural tendency and expected that studies commonly focusing on one topic highlight the importance of the particular topic

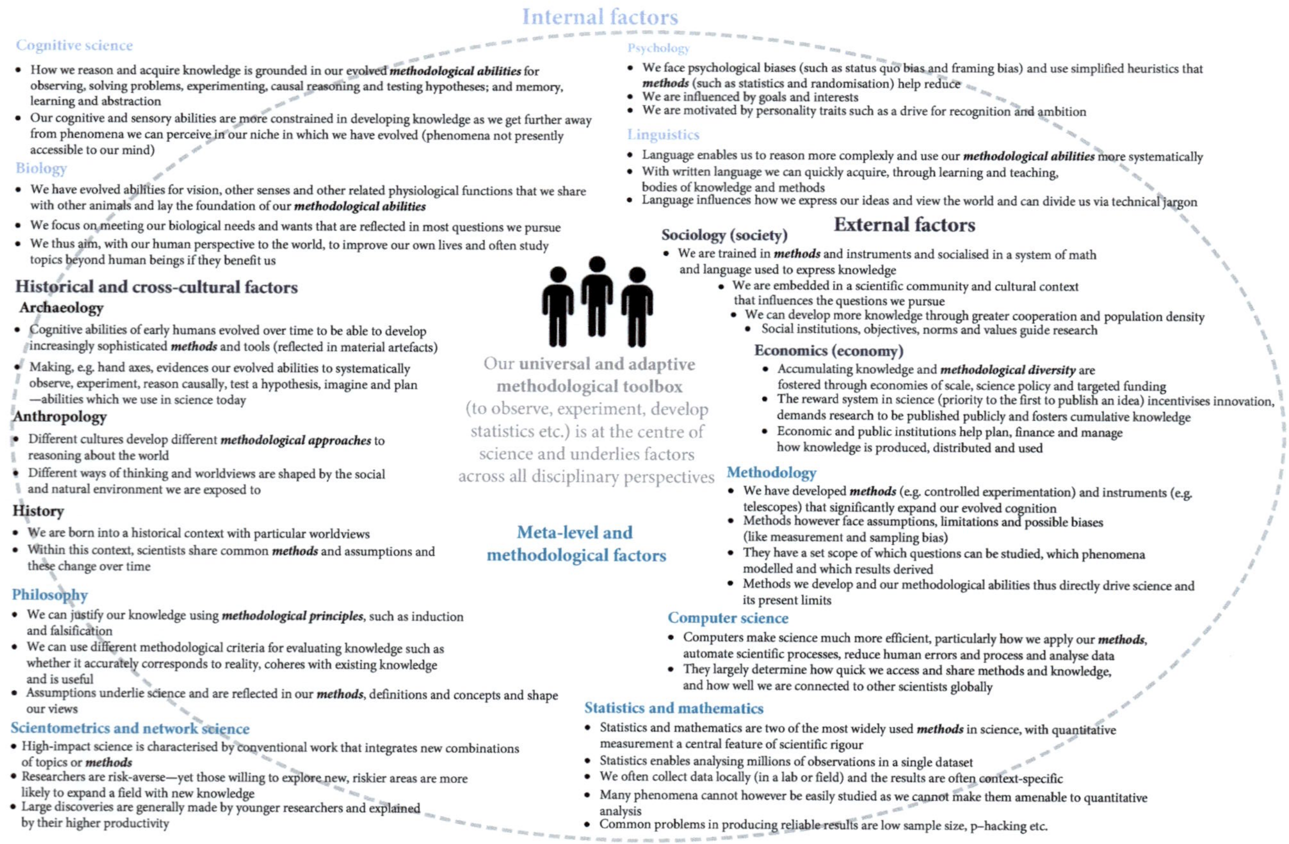

Figure 16.2 The origins, foundations and limits of science: the range of interconnected factors, and particularly our methodological toolbox, that enable and constrain science.

Note: Other influencing factors can be included for example into economics of science (such as the political usefulness of research) or into philosophy of science (such as ethical issues limiting what topics we can research and what experiments we can conduct). Regulatory bodies for instance place limitations on research related to human cloning, aspects of gene editing and technologies fostering climate change. We need to view the classification across the four areas loosely, with connections taking place across factors and areas.

they study. However, when we instead adopt a unified approach and integrated study, we can better evaluate the range of factors against each other, assessing the role of methods and instruments developed to be able to make scientific discoveries, the role of the traits of individual discoverers, the role of basic factors like collaboration and public funding, and so on. It is about assessing the importance of the different factors simultaneously and better explaining how we drive new advances.

A good analogy of the central challenge facing this field is that biologists, historians and archaeologists can study different parts of an ecological system and each provides valuable insights. But to understand the whole ecosystem it is necessary for the different researchers—and in addition for ecologists—to study the ecosystem in its entirety, from an integrated perspective. The research focus of science of science has, to date, been like studying ecology from different perspectives but not from ecology itself. *The critical missing piece in science of science has been for researchers to study the scientific ecosystem holistically, from an integrated perspective like ecologists do.* There is another way we can think about this central disparity in the field: the common approach of understanding science, by studying an aspect of science from one disciplinary perspective, is like trying to explain an ecosystem by only studying trees, or the human body by only studying cells. We acquire much knowledge, but that knowledge remains highly incomplete. To understand science, we require adopting a holistic-disciplinary approach and methods, just as we require doing so to understand other multidimensional phenomena, such as the environment and climate change (as illustrated in Chapter 1), the replication crisis in science (Chapter 13), health pandemics, poverty, and so on.

Overall, there is no consensus among researchers on which proposed explanation of what drives science is best and most accurate. Scientific consensus is however a central feature of science and advancement in other scientific disciplines, and for complex phenomena like science, consensus requires integration. To summarise, one central advantage of integrating the range of influencing factors in a single study is that it enables assessing the interconnected factors in a comparative way. This has not yet been possible in existing studies focusing on an individual issue or field, and without an understanding of the broader range of issues and fields and how they are connected. The trend in science towards specialisation in a single discipline has generated much specialised knowledge. But it constrains us in understanding a phenomenon as complex and diverse as science when not combined with a meta-approach that pulls the pieces together and provides an overall picture of science. The other central advantage of integrating the various factors is that it enables us to uncover which are most important and what central factor is shared in common across the different fields. In doing so, we find here that our methodological toolbox underpins the different factors across all disciplinary perspectives and is the only factor that does so. In Figure 16.2 we highlight the *methodological features* (in italics) in each of the subfields. This shift in focus to methods is central to the integration of the field of science of science. The degree to which other factors influence our scientific advances and discoveries varies depending on the phenomenon we study. This integrated approach enables us to develop a coherent understanding of science and grounds the new-methods-drive-science theory: *no factor plays as foundational and ubiquitous a role in understanding the origins, foundations and limits of science as new scientific methods and instruments we develop using our mind's methodological abilities* (as each field in Figure 16.2 illustrates).

No other factor influencing science is relevant in all fields—with for example linguistics and archaeology providing little, if any, insights into commonly mentioned factors like scientific funding, incentive structures, the scientific community and new theories. In ground-breaking scientific publications, we observe that teams can be small or large, low or high funded, young or old, at low and top ranked universities, or interdisciplinary or not. Money, collaborations and a research community are basic factors that foster science, but alone are not enough to break new ground. The central finding is that our universal and adaptive methodological toolbox (that enables observing, solving problems, developing microscopes etc.) is the main mechanism through which we *directly* drive science. All scientific advances and discoveries have been made using our methodological toolbox in new ways.(64)

We can better understand science in light of the integrated *new-methods-drive-science theory* presented here. This theory:

- places us, and the methods we develop using our mind, at the centre of studying science;
- integrates evidence of the abilities and conditions that have enabled us to develop science (biological, cognitive, social and methodological), the abilities and conditions shaping the scope of science (including, in addition, historical, economic etc.) and, most importantly, the abilities and conditions allowing us to expand the present limits of science (mainly methodological and instrumental but also cognitive, sensory and social);
- combines thus insights into the origins and foundations of science (especially our evolved methodological abilities of the mind and the methods

we develop) with insights into the present boundaries of science and how to push them (especially addressing our methodological constraints);
- pools together the range of methods used to study science to provide an integrated explanation that is consistent and better grounded in what is already known across disciplines (Chapters 2–15; Figure 16.2).

Methods are deeply embedded in our broader evolutionary, natural and social context. Because our methods are at the centre of how we do science and make discoveries, understanding the foundation of the methods we develop and use is at the centre of science of science. Our methodological abilities and scientific tools account for the foundation and different floors of our tower of science. The fundamental importance of our methods is evident for all aspects of science: conducting, evaluating and advancing science and also understanding science. It is thus surprising that among all papers studying science, such a small fraction analyse the nature of methods and instruments, their constraints and how to improve them (Figure 1.1).

This account of science explains how our methods and instruments and our human mind used to develop them set the scope within which we are able to develop knowledge and science. Then, beyond nature and our cognitive and methodological limitations in understanding nature, influences that are economic, social, historical and the like, even if often less direct and important, also shape the content and scope of the knowledge we create. Yet we cannot just focus on scientific methods, as we would otherwise be partially committing the mistake that traditional subfields make in focusing on one aspect of science—though on one integrated aspect of science across all fields (methods) instead of an isolated aspect of science in just one field (social norms, citation patterns and so forth).

In Chapters 2–15 we have synthesised the evidence and contributions of the different factors to our understanding of science. In light of this synthesis, we can depict the level of scope that a given factor has in explaining science, and the direct influence we have on that factor in shaping science (Table 16.1). These are the two criteria used to assess each factor. Some factors (and fields) mainly only help explain and shape the foundations of science (historical, anthropological, archaeological and linguistic factors) and others can also do so for the limits of science (biological, psychological and philosophical factors). However, we cannot directly influence them to promote science. These factors are thus classified as having lower scope to explain and influence

Table 16.1 Scope of factors/fields in explaining and shaping the foundations, limits and advancement of science.

Factors/fields	Explanatory scope: factors/fields explaining the foundations of science	limits of science	advancement of science	Direct influence: level of influence we have on a factor in shaping science
History **Anthropology** **Archaeology** **Linguistics**	X			Low
Biology **Psychology** **Philosophy**	X	X		Low
Economics (economy) **Sociology (society)** **Scientometrics/Network science**	X	X	X	Medium
Statistics/Math **Computer science** **Cognitive science** **Methodology**	X	X	X	Medium–High

Note: Cognitive science broadly covers our evolved cognitive abilities and constraints (related to observation, memory and abstraction), while psychology narrowly covers psychological biases and personality traits, as previously outlined.

science. Other factors (and fields), such as the economy, society and scientometric features, help explain and shape the foundations, limits and advancement of science. And we can partially influence them to foster science. These factors thus have medium scope to explain and influence science. Finally, other factors (and fields), such as methodology, statistics/mathematics, computer technology and cognition, also help explain and shape the foundations, limits and advancement of science. And we can most directly influence them to make new scientific advances. These factors thus have high scope to explain and directly influence science (Table 16.1). Our scientific methods including instruments are the only factor that underpins all 14 factors and fields (Figure 16.2) and that we are most directly able to influence to do and advance science.

17

Science of Science: An Integrated Field Grounded in the New-Methods-Drive-Science Theory

We offer here a foundation for the integrated field of science of science that studies science, and its foundations and limits, by combining methods and evidence from across the sciences (Chapter 1). In Figures 1.2 and 16.2 we outline what the unified field of science of science can look like. The field needs to be approached just like other fields of science: evaluating evidence cross-disciplinarily for consistency and coherence, applying methodologies comparatively and studying experimentally. To establish the field, some scientists of science and metascientists would need to be formally educated and trained, just as statisticians, psychologists and biologists are. We should not assume that some researchers across fields, who happen to develop an inclination for foundational and methodological questions of science, will come across such questions and find the time and resources needed to address them. Training researchers in science of science and methodology of science would help equip them with the necessary interdisciplinary and methodological skills to address present shortcomings in the field. While some scientists of science may focus on meta-level questions about science and scientific methods in general, others may focus on more specialised subfields such as economics of science or biology of science but do so in an integrated way coherent with what is already known in other subfields rather than in isolation.

Establishing the field of science of science requires providing not only an empirical foundation (Chapters 2–15) but also a theoretical foundation for understanding science. The new-methods-drive-science theory presented here can provide a unifying theory and foundation for the field that is grounded in the powerful role of scientific methods which is the common thread among this scientific community. The theory can integrate and unify the disparate fields studying science as our methods and instruments are connected to all features of science (Figure 16.2). Our evolved methodological

Science of Science. Alexander Krauss, Oxford University Press. © Alexander Krauss (2024).
DOI: 10.1093/9780198937401.003.0018

abilities of the mind (our *universal* methodological toolbox) and sophisticated methods and instruments we develop using our mind (our *adaptive* methodological toolbox) are the main mechanism that directly enables us to develop knowledge and science (Chapters 2–15). Our tools allow us to do science and also set the present limits of what science we are able to do. The theory describes how our methods have driven the origins, foundations and present limits of science.

The new-methods-drive-science theory explains how we advance science by developing new methods or refining existing methods that expand our present cognitive, sensory and methodological reach to the world. New methods and instruments we create—such as novel statistical techniques, X-ray methods and telescopes—enable us to make new breakthroughs by reducing our present constraints to studying the world in new ways. In contrast, existing leading (competing) accounts of science are outlined throughout the book—for example in the history of science by Kuhn who argued that science goes through paradigm shifts in theories;[1,32] in scientometrics in which scientists argue that career trajectories, team collaboration, research output and networks of scientists are the central parameters driving science;[4,5,9,35,37,38] in the sociology of science by Merton[130] and Latour and Woolgar[10] who highlight the central role of social factors shaping science; and so on.

In describing the new-methods-drive-science theory, we define the central terms here. Science is the study of the natural and social world by using our cognitive abilities (including observation, experimentation and problem solving) and the methods and instruments we develop (including statistical techniques and algebra, and particle accelerators and electrophoresis) with the aim of describing, explaining, predicting and controlling phenomena. Scientific methods are systematic techniques and scientific instruments are systematic tools used for scientific research and which are generalisable. In general, if more scientific methods are created, then more scientific progress will be achieved. The theory predicts that scientific progress will be brought about by generating novel methods, and the theory can be directly tested. And the theory is confirmed in a companion study assessing over 750 major scientific discoveries.[64] An assumption of the theory is that basic factors are in place, including our cognitive abilities and a minimal level of funding and collaboration to generate methods and tools.

The theory connects our new scientific tools to scientific progress. For they are what allow us to observe farther, process information better and measure phenomena more precisely, providing new perspectives to the world. Our methods and instruments are how we experiment with and control different phenomena in the world and expand our scientific scope. They determine how we design, implement and analyse studies and how we define and gather evidence. In science we require measuring, conducting experiments and analysing data on what we study. And to measure, conduct experiments and analyse data we require methods and instruments. Without them, we are not able to describe and explain most complexity in the world. The new-methods-drive-science theory is thus a general theory of scientific advancement through methodological advancement.

Yet where does the demand for developing new methods and instruments arise? The demand in contemporary science is driven by scientists who require a better method or instrument to better study a phenomenon. Chromatography, the Geiger-Müller counter, the spectroscope and electrophoresis for example has each been applied to make at least a dozen Nobel-Prize-winning discoveries, and scientists developed each of them.[(138)] How we advance science can be categorised in five main stages. The *methodological constraint* needs to first be identified, as researchers generally run into a practical problem they cannot solve when using a method or instrument to study a phenomenon. Next, *methodological scanning* is the process of scanning one's own field and related fields for a method or tool that can address the given constraint—and if not found, moving to the next stage. *Methodological conceptualisation* is the process of conceiving the new method or instrument designed to address the methodological constraint we face. *Methodological development* is the process of acquiring the needed resources and creating the new method or instrument that enables us to measure and observe the world in new ways. And *new-methods-driven discoveries* is the final process of applying the new method or instrument to make a new breakthrough.

Within this new-methods-drive-science framework, we can begin to strategically plan and target efforts to create scientific tools that enable us to expand the scientific frontier. There are two central ways for us to increase the speed at which we develop new breakthroughs. One way is directing more attention to expanding existing methods and instruments, recombining them in entirely new ways and creating completely new ones that allow us to make scientific advances. The other way is using scientific tools available in other fields to address questions in novel ways and also sharing knowledge of new scientific tools across fields immediately once they are created. To do so, leading journals and institutions will need to incentivise researchers to develop and publish methodological innovations. For some methods and instruments are often not used in other fields for decades, such as RCTs that were applied in medicine long before psychologists and economists began applying the method that revolutionised their fields.

A central point in this book that goes against the common view in science and science policy is that science is and should be question driven. Scientific publications are generally structured around a research question. Novel methods and tools of science have however often led to scientific breakthroughs by conducting exploratory research using them—independent of any existing or not-yet-formulated questions. Making scientific advances can be fostered by focusing our attention not just on question- and hypothesis-driven science but also on methods-driven science that is exploratory (without any predefined question). Academic journals, grant committees and university hiring panels need to equally recognise the development of methodological advances as central to doing and advancing science. The hope of this book is to help spark such a methodological shift in how we view, carry out and reward research.

Our methodological toolbox, through collective mind–method synergies, has been and will continue to be the main mechanism we have to address the big questions and challenges we face. These range from developing new technologies that shape contemporary life, to extending the way we make evidence-based biomedical, social and environmental decisions. So how can we improve science and push the boundaries of science? The answer to this question helps us better understand the science of the future—and the answer lies in understanding the power of our methodological toolbox and in taking steps to refine it. The answer involves more effectively tapping our abilities to use and improve our methodological toolbox in innovative ways—the topic of the remaining chapters.

A final important realisation and implication here is that some of our leading scientific methods, from statistical methods to RCTs, could have turned out differently from how they did. And they will turn out differently from how they currently are as we continue developing them over time. Placebos, double-blinding, p-values and the like are not inevitable. Other methodological solutions to the problems they aim to address are conceivable when conducting controlled experimentation.[(42)] Once we become aware of this fact, it opens our minds to the undetermined power of the methods we create and the possibility of developing a range of previously unimaginable methods to address new problems and challenges we face. This would enable us to continually expand and reshape our tower of science. And it could mark the beginning of a methodological revolution in science, and enable us to continually push back the present limits of science.

18

The Limits of Science: An Overview

In the next three chapters we will now pull together the different evidence to outline the present limits of science and, most importantly, how to push those limits and expand our research frontier. Are we approaching the boundaries of science? Can we continue advancing at the periphery of space, matter and complex ecological, biological and economic systems? Are we close to reaching the boundaries in fields like fundamental physics and brain science? These are fundamental questions about science we have not yet addressed. We know that the universe came into existence about 14 billion years ago, that fundamental forces govern physical reality (gravity, electromagnetism, the strong force and the weak force), that the periodic table of elements represents the chemical elements that make up the world, that the universe and life evolve, and that DNA carries genetic information needed for biological organisms to develop, function and reproduce. These make up essential pillars of science, so they would unlikely be substituted by completely different breakthroughs and theories that are as extraordinary.

Are we thus nearing the boundaries of science and significant breakthroughs? Since the turn of the millennium, many groundbreaking discoveries have been made, such as the Nobel-Prize-winning discoveries of CRISPR gene editing in 2012,[(245)] the Higgs particle also in 2012 and the existence of gravitational waves in 2015,[(246)] among other major discoveries that did not win the Nobel Prize such as the mapping of the human genome in 2004. These recent discoveries redefined the frontiers of genetics, physics and astronomy and it does not appear that science will stop expanding soon. But how did we make these discoveries? Each new breakthrough advancement that pushed the frontier has been propelled by a newly developed methodology or instrument: in these cases, a new method of differential RNA sequencing developed in 2010, the large hadron collider (particle accelerator) in 2008, the upgraded LIGO laser interferometer in 2015 and an improved genome mapping technique in 2001, respectively. To go beyond our present scientific horizons, we must comprehend their present limits and the particular constraints we need to tackle to extend those limits. We need to specifically understand the limits of our current methods and instruments that

Science of Science. Alexander Krauss, Oxford University Press. © Alexander Krauss (2024).
DOI: 10.1093/9780198937401.003.0019

enable us to study, measure and perceive the world and that shape the limits of our current knowledge. For they enable us to gauge our scientific progress and when we reach the borders of how we presently study the world.

At the forefront of research, many mysteries exist that we have not yet solved: understanding the foundations of mathematics; the fundamental nature of consciousness, memory and time; identifying new mechanisms for extending our lifespan and mitigating climate change; comprehending what our universe is comprised of, and how stars, planets and human cooperation evolve,[(247)] how to eradicate cancer, and how to best use artificial intelligence and machine learning to advance science and society.

What our present limits of science are and how to address those limits are among the most important questions we can ask. While biology and medicine have made vast strides in reducing disease and improving life expectancy, many illnesses persist. While physics and chemistry have brought about vast technological and industrial advancements that shape contemporary society, they have also fostered problems such as climate change, environmental degradation, warfare and mental health issues related to dependencies on electronic devices, which we have not yet solved. The scientific community does not generally investigate our scientific boundaries as they are seen as difficult to study systematically, and thus a comprehensive and interdisciplinary analysis does not yet exist.

Researchers have generally studied *what* particular phenomena we are not yet able to study—not *how* we push our boundaries of science to be able to study new phenomena in the world. *Science* published a special issue that outlines 25 large outstanding questions facing science.[(247)] Researchers studying the boundaries of science have in turn investigated the topic taking a perspective from physics,[(248,249,250)] mathematics[(249,251,252)] or particular fields,[(253)] from the human mind[(254,255,256)] and philosophical aspects about scientific laws and scientific induction.[(220,252,257,258,259)] In a book that aims to study the limits of science, Gleiser surveys the history of physics and aims 'to illuminate a variety of scientific and philosophical viewpoints' largely by exploring conceptual shifts in theories of physical reality and their limits.[(248)] In *The End of Science*, a highly influential and cited book on the boundaries of science, Horgan pessimistically argues that science may be nearing its end as the big questions have nearly all been addressed.[(253)] But given the recent major scientific advances and big open questions outlined above, we can see that this is not the case. An integrated explanation of the present boundaries of science yet remains elusive. Understanding our scientific boundaries and how we can extend them requires however combining methods and evidence from different fields as they provide different pieces to the overall narrative.

We will explain how our methods and instruments and our mind used to develop them set the present limits of what we can know and what is possible in science—and economic, social, historical and spatio-temporal influences help shape what we study within those limits. Researchers inevitably conduct science applying a given method or instrument and our cognitive abilities, and do so from within our spatio-temporal context. Developing methods and instruments enables us to continually push back those present limits and is commonly the factor we can most directly and strongly influence to do so. By creating new methods and instruments, we can reduce our present cognitive, methodological and social constraints to investigating the world and vastly expand our scope and understanding of the world. Science reaches its limits in explaining reality as it reaches the limits of the tools of science we have created thus far which enable perceiving and representing nature. Expanding science takes place at the pace at which we design new methods and instruments that enable us to do science in new ways. Extending our bodies of knowledge represents the shifting scientific borders reshaped by extending our methodological toolbox. Methodological growth is necessary for major scientific progress.

Viewing new methods and instruments we develop as the main mechanism of how we advance science provides a new framework for addressing questions about expanding the limits of science. Understanding how we advance our frontiers by developing new tools and how cognitive, social and economic factors foster those new tools is the best way we have to understand what shapes the scope of science and its limits and, most importantly, how we can extend those limits. The paradox is that by understanding what constrains us we can best overcome those constraints and continue to advance science. This new-methods-drive-science mechanism here has wide applicability in advancing science across fields.

In this chapter we first explore historically how we have extended the borders of science and we then provide a general conceptual description of the scope and limits of science. In the next chapter we illustrate the particularly powerful role of the present boundaries of our methods and instruments shaping the present boundaries of science. And in the following chapter we describe how we can make advances at the frontier quicker by describing the steps to extend our scientific tools to study the world in novel ways. We also discuss whether there are pre-established boundaries to some domains of knowledge and we outline pathways of the future of science that we can take.

Scientific progress over time: persistently pushing back the borders of science

In the early 16th century, most regarded the physical universe as geocentric and finite. However, advancements in scientific methodologies and instruments emerged during the 17th century, including calculus, the telescope, statistics and the barometer. These innovations facilitated novel approaches to investigating and theorising about the world, ultimately leading to the widespread acceptance of a heliocentric and infinite physical world by the end of the century. In the world of medicine, a systematic understanding of how diseases, treatments and their causes were connected was lacking in the mid-19th century. However, by the early 20th century, the creation of systematic controlled trials and the methods of randomisation, blinding and placebos, which were combined with statistics, revolutionised experimental practices. These methodological enhancements significantly improved our ability to assess the causal effects of medical interventions, vastly enhancing our well-being and lifespan. How we give rise to such new understanding of the world reflects a general principle across science: *developing new methods and instruments of science is how we expand the research frontier in new ways.*

We have stretched the periphery of science continuously throughout history. To reduce our limited vision, we created instruments such as the microscope in 1590 and then the telescope in 1608 that revolutionised science.(260) The advent of microscopes transformed our understanding of human disease, offering unprecedented insight into viruses, bacteria and microorganisms. The development of telescopes obliged us to reassess our knowledge about planets, our solar system, the size of the universe and our place in it. We thus reshaped the boundaries of disciplines like biology and physics which were then no longer set by the bounds of our sensory abilities but by the bounds of these new tools. Such tools were refined and very powerful up to the 19th century until we hit the blurred boundaries of their resolving power and thus again the boundaries of science in different domains. In the decades up to the 1930s, scientists could perceive many phenomena but not others, despite awareness of their existence. However, the creation of the electron microscope in 1933 marked a pivotal moment, enabling researchers to explore previously unseen realms such as living cells, large molecules and crystals.(261) The electron microscope revolutionised multiple disciplines and forced a redefinition of their boundaries, which were once again set by this groundbreaking tool.

To reduce our cognitive constraints, we developed randomised controlled experimentation in 1948 that reduces biases in the scientific process and enables us to better assess how effective our best medical treatments and public policies are. This method redefined the way we evaluate and understand the effectiveness of our medications and economic policies.[42] We thereby reshaped the boundaries of science across the biomedical, behavioural and social sciences. The present bounds of science are largely synonymous with the current bounds of our best tools we have developed. In this sense, only by constructing the telescope for example was Galileo able to discover Jupiter's moons in 1610. Only by developing calculus was Newton able to describe the laws of motion in 1687. Only by creating X-ray diffraction methods were Franklin, Crick and Watson able to detect the double helix structure of DNA in 1953. Such discoveries would otherwise not have been possible.

Throughout history, the advancement of science has predominantly been attributed to the efficacy of the tools we have created. This fundamental fact has been largely overlooked, leading to the absence of a comprehensive theory of science outlining the role of methods in driving scientific progress and defining its limits.[220,247,248,249,250,251,252,253,255,259] The scientific periphery has been extended without a comprehension of the pivotal role played by novel methods and tools. We have redrawn the limits of science in an ad-hoc way, by individual researchers who happen to expand a given tool that enables better studying a given phenomenon.

New tools have thus driven our expanding scope of the world in a cumulative process. In the 17th century, the emergence of six groundbreaking methods and instruments triggered most scientific breakthroughs at the time: the microscope developed in 1590, telescope in 1608, barometer in 1643, air pump in 1659, statistics in 1663 and calculus in 1675. The leading scholars, Galileo, Hooke, Boyle, Newton and their contemporaries, each leveraged one or more of these new methodological innovations to expand our understanding in astronomy, biology, physiology, pneumatics, mechanics and optics. These tools and their refinements continue to occupy an essential role in scientific research. Scientific progress is thus not driven by chance but guided by developing innovative scientific tools (Figure 18.1).

As we expand science, pushing the frontier has required increasingly sophisticated methods and knowledge over the centuries. Hooke could discover cells in 1665 using a novel low-power microscope. Newton could explain gravitation in 1687 using mathematical methods and direct observation.[223] Today, to contribute to our existing body of knowledge on cells or gravity we need to first learn how to apply the needed complex methods and instruments. We must also study the existing body of knowledge on cells (by Flemming, Claude and others) or existing theories of gravity (by Newton, Einstein, quantum theorists and others) before being able to

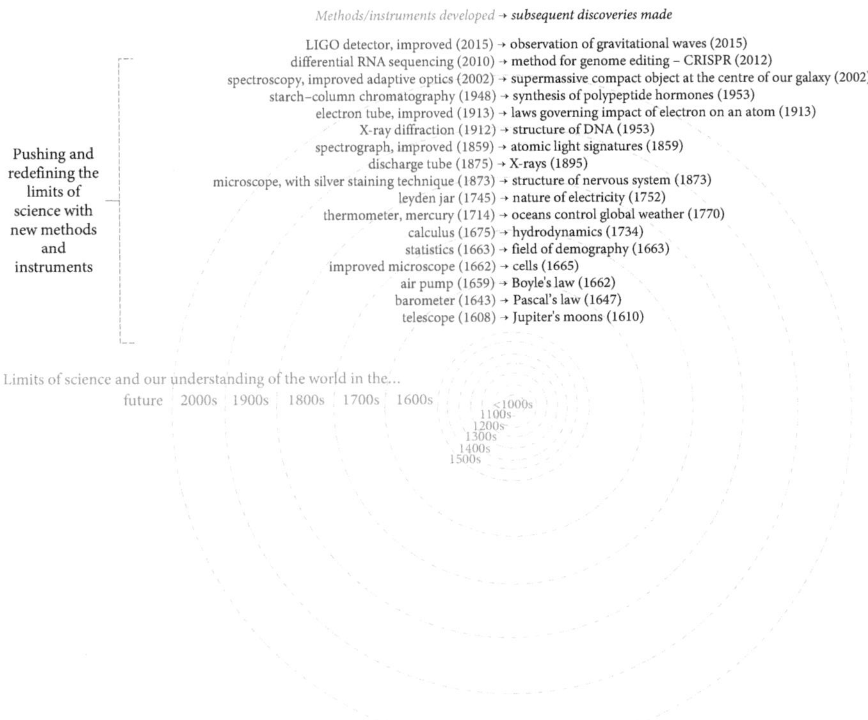

Figure 18.1 How we comprehend the world expands at the rate we advance new scientific methods and instruments.

identify ways to expand them. Galileo could identify the moons of Jupiter in 1610 using the recently developed telescope, but to establish which chemical compounds make up the moons of Jupiter and how they came into existence requires first developing highly advanced telescopes and spectrometers. Without such methods, instruments and knowledge, we would not be able to push the research frontier and we would not know we have without first acquiring the existing knowledge. Many *low-hanging-fruit* discoveries have been made using two paradigmatic instruments—microscopes and telescopes—by simply pointing the new instrument (with increasingly greater power) at phenomena around us, without searching for them. Yet such discoveries have become less common over the past centuries. For each generation it thus takes us longer to reach the frontier.[cf.(171)]

The success of science is commonly viewed as culminating in the great theories in physics of relativity and quantum mechanics that explain the world on both the smallest and largest scales. It is reflected in the great discoveries in biology of the structure of DNA and the mechanisms of evolution that explain the living world and the secrets of life.[cf.(262)] Such discoveries have generally given scientists the impression that there may not be big missing pieces to understanding the world. We cannot however understand how science will progress in the future by just viewing existing discoveries. We need to understand what drives discoveries. Here we illustrate that major breakthroughs are mainly driven by methodologically inclined scientists and inventors—those who design our electron microscopes, radio telescopes, spectroscopes, contemporary statistics, X-ray diffraction methods and a vast array of methods that have extended our lens to the world. These are the heroes of science who make studying the world in new ways possible.

Despite greatly expanding our boundaries of science, many unsolved mysteries remain that inspire us: Is there other intelligent life in the universe? Where do black holes come from? What are the causes of many mental disorders?[(247)] What drove the cognitive revolution in early humans? How much of mathematics is explained by nature or culture? How do our brains' electrochemical impulses transform into our cognition and emotions? How can we reconcile gravity with electromagnetism? How can we best change human behaviour to mitigate environmental degradation? May technological intelligence one day exceed human intelligence (that is, is so-called singularity attainable)? And so on. The unknown fascinates us. Providing answers to fundamental questions also generally gives rise to new, not-yet-conceived questions in an iterative process at the research frontier. The key factors constraining and enabling us to address such fundamental questions are our methods and mind, and also our social, biological and spatio-temporal influences.

The scope and limits of science: the border between the known and the unknown

What and how we perceive the world is invariably shaped by the scope of our methods and cognitive and sensory abilities—that is, our methodological toolbox—but also often by our social, economic and historical influences as well as our human needs and objectives (Figure 18.2). Together, they set the boundaries within which we are able to observe, process and understand phenomena in nature and society. We then push the present limits of science by developing tools that reduce our cognitive, methodological, social and other human constraints. Methods and instruments are a necessary condition for major scientific advances. Again, among these different factors, developing

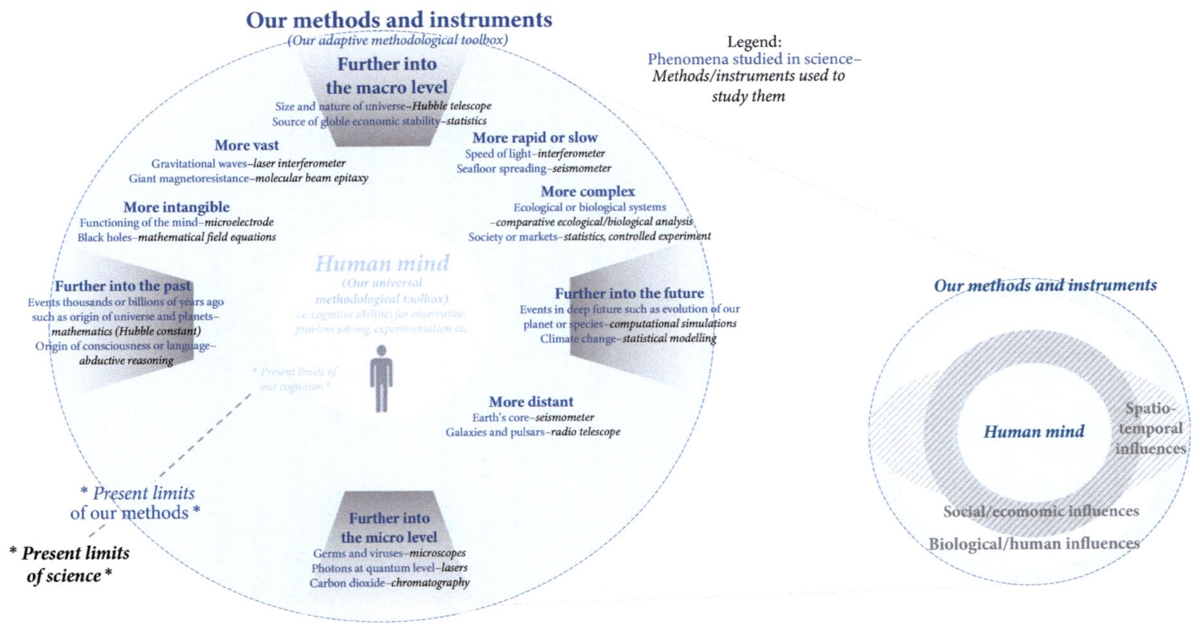

Figure 18.2 The boundaries of our mind and methods (our universal and adaptive methodological toolbox) shape the present boundaries of science.

Examples are provided for each category as illustrations, with the phenomena also studied using other methods and instruments. Phenomena can fit into more than one category, with for example ecological systems being macro-level phenomena with greater levels of complexity.

methodological innovations is the feature of science that we can most directly influence.

Beyond our methods and what our mind is methodologically capable of developing, we have no alternative way to broaden our scope to the world. Science and its boundaries are largely fixed in a circle that begins and ends with our tools. But we continually reshift the lines of that circle by inventing better tools. Comprehending the bounds of science is thus contingent on recognising that we humans carry out science, and we face cognitive and methodological constraints to doing science that delineate our present bounds at any given point. Pushing the research frontier is about reducing these constraints.

Science and knowledge can be understood on a spectrum: with knowledge of phenomena that are directly observable at one end of the spectrum—such as flora, fauna and our habitat. This is how we have acquired much of our knowledge in fields like botany, anatomy and palaeontology. Then there is knowledge of phenomena that lie beyond the directly observable conditions in which our mind and senses have evolved but that we can access using our scientific instruments and methods—such as microscopic cells, quanta, galaxies, statistical probabilities of diseases and global pandemics. And, at the other end of the spectrum, there are theories presently developed about phenomena that lie beyond those conditions and we lack sufficient empirical evidence of—such as the size of the universe, the historical origin of life, string theory and singularity (Figure 18.3). We cannot access, with our mind or methods, such phenomena as we move further away from our cognitive and methodological niche.

As the present boundaries of science are most strongly and directly shaped by the boundaries of scientific methods that we develop using our mind, we turn to this topic in greater detail in the next chapter. There we also demarcate when we, given the limits of our methods and mind, cannot verify our scientific theories and thus reach the limits of science.

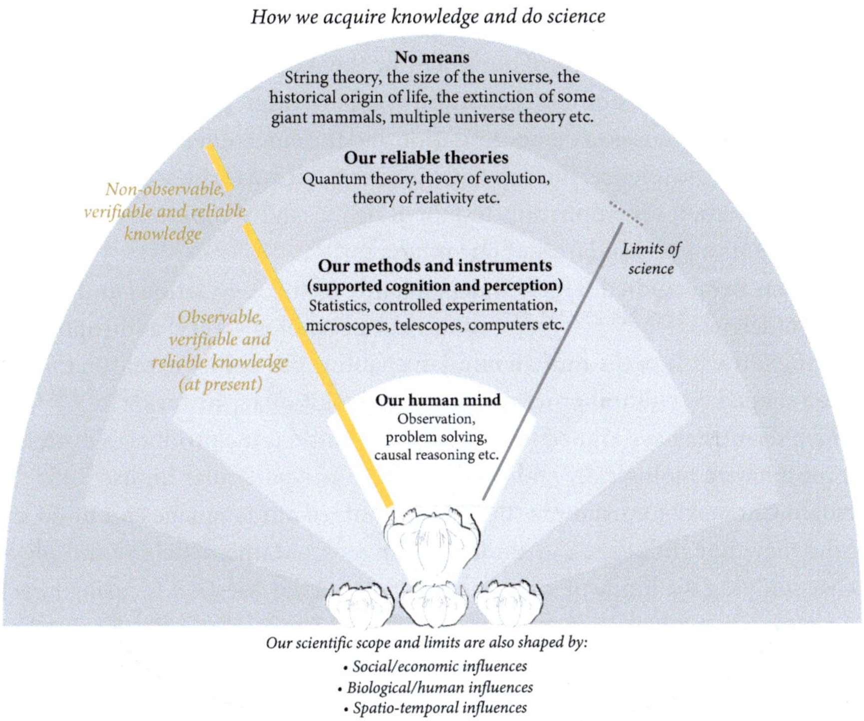

Figure 18.3 The scope and limits of science.

19

The Limits of Science: Grounded in the Boundaries of Our Methods and Mind

We still poorly understand where we draw the boundaries of science and when we reach the present limits of our scientific theories about the world. Influential theories about the size of the universe, superstrings, the historical origin of life, the general causes of democratisation and the emergence of our consciousness exist across science. But can these influential theories be scientifically reliable if we cannot yet empirically test them rigorously? We know a lot about the foundations and boundaries of our mind,[21,24,50,74,81,87,102,104,106,109] but the topic has been studied largely independently of the foundations and boundaries of science.[248,249,250,251,252] Biologists, psychologists and anthropologists highlight the role of the human mind in enabling our ability to reason that has been shaped by cultural processes over our evolutionary history.[21,52,88] Cognitive scientists have argued that because our mind is the product of evolution it must have a biologically endowed structure and particular limits.[254,255,256] Noam Chomsky for example studies the limits of our language and mind and how they may influence our understanding.[256] Mathematicians and physicists studying the limits of science have explored the question focusing largely on areas in mathematics, physics and their history.[248,249,250] Researchers have not yet explored the foundations and limits of our mind and how they specifically shape the foundations and present limits of our knowledge and science, by integrating evidence of the evolutionary origins of our mind, and the boundaries of science and scientific methods we develop using our mind. We do so here to explain how our cognitive, sensory, methodological and instrumental capacities and constraints drive the present limits of science, influencing the theories about the world we are able to develop and test and those we are not yet able to. We explain what types of phenomena are accessible to scientific investigation and what phenomena are presently not.

Our cognitive, sensory, methodological and instrumental abilities generally become more constrained and less reliable in acquiring knowledge and doing science as we get further away from studying phenomena that we can perceive within our environmental and cultural niche in which our mind and

Science of Science. Alexander Krauss, Oxford University Press. © Alexander Krauss (2024).
DOI: 10.1093/9780198937401.003.0020

senses have evolved. This is especially the case as scientists develop theories about phenomena like the size of the universe, dark matter and the general causes of democratisation that are far beyond that niche. *The central argument here is that we reach the present limits of science, and what science itself is, when our theories involve phenomena that are not observable and thus the theories are not verifiable and empirically reliable using our mind, methods and instruments. This is called here the OVER criterion of science (for observability, verifiability and empirical reliability).* Thus, if real phenomena in the world are not directly or indirectly *observable* using our mind, methods or instruments, then we lack sufficient evidence to develop theories about them that are *verifiable* and *empirically reliable*, which is when we reach the present limits of science. These are the features of phenomena that fall beyond our sensory range that we are not able to perceive and are not presently accessible to our mind, including the methods we develop using our mind–from x-ray devices and lasers to computational techniques. They reflect the conditions in which we hit the boundaries of what we can reliably study. It is defined as the point at which phenomena are disconnected from our direct perception (unaided, using our cognition) and from our indirect perception (aided, using methods and instruments we develop using our cognition). And we cannot yet access such phenomena in the world. The unique feature of this account of science is the focus on and evidence provided by the evolution of our mind and especially the present boundaries of our mind, methods and instruments. The OVER criterion of science thus reflects the particular features of phenomena that, given the limitations of our mind and methods, reach the boundaries of our scientific theories and science—as we outline in the chapter.

This explanation provides a new foundation for grounding science and its central evaluation criteria of empirical testability, verifiability and reliability. These are the essential features for explaining and understanding the present boundaries of science: whether we view evidence or a theory about the world as reliable depends on whether our evolved cognitive system (including the tools we create using it) is able to observe the given phenomenon and test and verify that evidence or theory. In this chapter we focus on *scientific theories* and their limits. A scientific theory aims to explain a phenomenon in the world;[(263,264)] to 'explain how or why something happens' in the natural or social world.[(265)cf.(266,267)] If our theories involve phenomena that do not meet the OVER criterion, if they are thus not observable, verifiable and empirically reliable, then we are not able to explain the phenomena and we reach the limits of our scientific theories.

In this chapter we first provide a novel empirical explanation for the present limits of science, namely how our human mind, methods and instruments constrain the theories we are presently able and not able to develop—that is, theories about non-OVER phenomena. We then provide insight into the origins and foundations of science's central evaluation criteria that are grounded in the OVER criterion of science. Overall, we outline the foundations and limits of scientific theories and thus science in light of the foundations and limits of our mind and the methods and instruments we develop using our mind. What emerges is the first account of a criterion of science that demarcates the foundations and limits of scientific theories and thus science which is grounded in evidence on the evolutionary origins of our mind and on the limitations of our cognitive and methodological abilities. We will then draw implications for how the OVER criterion of science can help clarify ongoing debates about influential scientific theories—such as on multiple universes and superstrings—that have not yet been verified.

The borders of science are shaped by our human mind and methods—our universal and adaptive methodological toolbox

While our mind and senses have evolved over human history largely to ensure our basic needs and survival within our observable niche of the world, by extension they have evolved in more recent history—especially over the past 70,000–100,000 years—in ways that enable us to formulate theories about many phenomena that are not directly observable. We have accurate theories about many phenomena we have direct (generally observational) experience of and which we can thus empirically test and verify. These include flora, fauna and habitats within our evolutionary niche. This is also the case for many phenomena that are not directly observable, such as in chemistry and biology, but which we can indirectly access and empirically test and verify using our developed methods and instruments. In science, attaining knowledge about real phenomena in the world is thus only possible for phenomena we can get into contact with using our senses, mind and the tools that our mind is capable of conceiving.

We face, in contrast, overwhelming constraints when it comes to formulating and reliably assessing theories about phenomena that do not fulfil the OVER criterion of science, such as multiple universes, the historical origin of life, dark matter, the origin of the moon and the evolution of conscious experience. For they go beyond our spatio-temporal niche accessible to us using our mind and methods. What such theories have in common is namely insufficient empirical data to prove them. The best we can do is make inferences to the best explanation by interpreting and extrapolating from (related) available evidence. These particular theories, if sufficient empirical evidence may arise in the future, may turn out to be correct. But without such evidence, we are not able to provide reliable and replicable knowledge. In many cases we cannot reconstruct a phenomenon or restart a historical process to observe how it has evolved.

The point here is not to isolate or criticise particular theories that are not yet or may not become verifiable. But it is to highlight that the features of such theories importantly illustrate when our knowledge cannot be reliable and where the border of science lies: a border drawn by the OVER criterion of science. While the present limits of science are dynamic and we can push them back as we develop new methods and evidence, we reach those limits, at any given point in time, when we formulate theories that are not yet testable, verifiable and empirically reliable (non-OVER theories).

In Figure 18.2 we outlined how we are able to study phenomena using our universal methodological toolbox, with abilities for observing, solving problems and experimenting that we are all born with. And how we are able to do so using our adaptive methodological toolbox, with statistics and X-ray methods that we collectively develop using these abilities and pass along within groups and scientific communities. Together, they are what enable us to acquire our vast bodies of knowledge about the world. Our universal methodological toolbox reflects these evolved methodological abilities of the mind and allows us to directly observe those parts of the world that our mind has largely adapted to (our human domain or niche). Using our universal methodological toolbox we can thus create sophisticated methods (our adaptive methodological toolbox) to broaden our scope to the world. This enables us to access and make sense of different phenomena (from chemical to immunological phenomena) that lie beyond the scope of our universal methodological toolbox and thus our human mind. We can think of the two as our core and extended methodological toolbox.

Our early ancestors were likely not able to explain mysteries such as lightning, shooting stars, fire and many diseases. Today, we understand them well thanks to the methods and instruments we have developed such as telescopes, spectrometers and controlled experimentation. But our present methods only enable us to partially understand phenomena like cancer, the origin of life, many functions of the brain and the global economy. And we poorly understand phenomena such as the size and nature of the universe and the evolution of our consciousness and language, as we have not yet created the methods and instruments needed to shed light on them. Methods and instruments we develop have enabled us to continually push back and redefine the present limits of our scientific theories and thus science, and have driven scientific, medical and technological advances.

It is with the present boundaries of the tools we have developed thus far where we reach the present boundaries of what we can observe, test and verify and thus what is reliable knowledge. Using our evolved mind we gather information through our vision, perception of temperature and sense of weight, time and speed. To stretch our sensory and cognitive abilities, we create instruments like particle accelerators, electron microscopes and chromatography that have vastly increased our knowledge of the world of atoms, microorganisms and chemical substances. To extend our cognitive abilities to process that information, we develop methods like statistics and controlled experimentation that have greatly improved our understanding of phenomena such as effective medical treatments and human behaviour. Methods and instruments help make phenomena that are not directly accessible to our senses indirectly accessible to us and allow us to better make sense of those parts of reality we can perceive directly with our bare senses. Think of what is needed to study the world with thousands or even millions of observations using regression analysis, for example of climate conditions or medical patients. We were not able to do so until we developed contemporary statistical methods and computers in the second half of the 20th century (Chapters 12 and 13). Combining statistical methods with computers revolutionised many fields of science, from experimental physics to medicine and psychology. For it made large-scale statistical analysis possible for the first time. This has been one of our greatest methodological revolutions. It has also helped reduce researcher biases through mechanical and automated processes.[30,42]

As we have seen in previous chapters, the purpose of developing a new method or instrument is to improve our understanding or solve a problem by reducing particular cognitive, sensory or methodological constraints. To this end, microscopic and telescopic tools enhance the scope of our retinas to reveal previously unimagined worlds of life within a drop of blood or water, and moons around planets. Our mind is also not generally able to control for, or understand well, the complex processes of many multivariate factors, observable or unobservable, operating together at different levels within biological, physical or social systems. Statistics enable us to better understand such complex systems (Chapter 13) as they allow us to reason inferentially with a vast amount of data and process and analyse those data in complex ways and over long periods of time. Our mind is not, without methods and instruments, able to do such tasks.

Methods and tools are not just means of expanding our mental abilities. They at times also represent a different mode of reasoning and theorising by providing a different (often quantitative) way to view phenomena and a different scope to the world. Our universal and adaptive methodological toolbox sets—at any given point in time—the present boundaries of what phenomena we are able to observe, measure, theorise about and understand, and what phenomena we are not yet able to. It makes up our world, our empirically verifiable world.

Our evolved, universal methodological toolbox reflects our senses and intuitions that are most well adapted and we are most comfortable with. Without developing highly complex tools, we would not have otherwise developed theories about most phenomena beyond our directly observable realm—theories about chemical compounds, our biosphere, infrared light and gravitational waves. To do so, only methods and instruments can extend our reach. Our basic perception and experience have been and remain at the centre of how we develop our understanding of the world.[cf.(107)] All scientific fields once began using only our universal methodological toolbox. And fields like botany, zoology and anthropology still mainly only use that toolbox, while other fields like chemistry, molecular biology and astrophysics mainly study phenomena using our adaptive methodological toolbox. Pushing the boundaries of science is about collectively stretching our cognitive and methodological resources. In a nutshell, that is the core of science and how we expand science.

We generally face more challenges in understanding phenomena the further we move away from our human domain—the further we look back or forward in time, the smaller and larger the phenomena we study, and so on (Figure 18.2). We reach the present boundaries of science when it is no longer possible to gather new data—or interpret and extrapolate from existing data in new ways. Quantum mechanics is a peculiar exception as it is not very intuitive but provides reliable predictions of the behaviour of atoms. In general, the reason why our theories often become less reliable as we move further away from our human domain is because we generally have less empirical evidence to back them up.[(34,107)] We access a part of the world concentrated in our human domain and accessible with our methodological toolbox. This is evident when we think about all potential knowledge of the world—particular phenomena at the micro and macro level (that presently lie beyond our cognitive, sensory and methodological abilities), with high levels of complexity (that presently lie beyond our cognitive and methodological abilities), over the span of history (that presently lie beyond our spatio-temporal means of data collection) and outside our anthropocentric viewpoint (that presently lie beyond our biological reach). When we study phenomena with increasing complexity, or with increasing depth at the micro or macro level, the present limits of science are generally shaped by human limits—our cognitive and methodological limits. Yet when we study phenomena further back or forward in time, the present limits of science are shaped by those human limits but also by greater difficulties in accessing and gathering data as we move further into the past or future.

Our methodological toolbox is thus the driving force of science and its present limits. This basic fact is evident given that species, including our own, do not evolve the cognitive and sensory abilities to be able to perceive phenomena beyond what is directly observable, such as quarks and other galaxies, and given that it is only through the tools we humans have recently created that we are able to observe and verify such phenomena. Yet we need to understand the present bounds of science loosely, changing and evolving as we expand and improve our adaptive methodological toolbox.

The present limits of science are grounded in the present limits of our methods and mind: observability, verifiability and empirical reliability

We can access and verify phenomena directly (such as plants and animals) and indirectly (such as molecules and bacteria). And our evidence and theories about both types of phenomena are more concrete, reliable and replicable (using multiple methods) than our theories about phenomena like the size of the universe and dark matter. For these are presently not penetrable with human sight or our instruments. We poorly understand such unobservable phenomena and we face difficulties in attempting to develop theories about them. This is partly explained by the basic fact that they are not the kinds of phenomena that our mind and sense organs have evolved to be able to perceive and process, or that the methods and instruments we create can deal with. In this sense, the reliability of our best theories and the lack of reliability of our non-verifiable theories are inseparably linked to our cognitive and methodological abilities. While we for example cannot directly observe the evolution of our human mind itself, we have many different forms of empirical evidence to test and verify it and thus also the theory of evolution. Anatomy describes how species share similar traits that were present in a common ancestor. Genetics demonstrates the degree to which species are related. Archaeology studies fossils to illustrate how species, including our own, change over time and how extinct species relate to living species and intermediate species.(268)

Theories reaching the present limits of science, such as on the fundamental structure of reality, the historical origin of life, dark matter and the size of the universe, can be called *unverifiables.* They are qualitatively of a different level of abstraction than theories that deal with not directly observable entities like molecules, cells and viruses but that we can, in contrast, indirectly observe using for example electron microscopes. With unverifiables—defined as phenomena that are disconnected from our direct or indirect perception—we do not have sufficient empirical evidence to verify theories about them. We are captivated by these deep questions concerning distant origins, vast complexity and fundamental structures, but addressing them eludes us given our current limitations in perception and experimentation. This is where science stands still and we reach the threshold of our knowledge. Theories like those about superstrings and multiple universes lack empirical evidence. Without good empirical evidence, theories do not stand the test of time. We can thus classify theories about different phenomena into one of three types (Table 19.1).

No way exists for us to perceive and verify phenomena in the world and deem them reliable except by using our mind and what the mind is methodologically capable of. It is thus a basic fact of science that our present limits of science are primarily defined by our present cognitive and methodological limits. Social organisation, including institutions and funding, can also influence aspects of the limits of science, such as financial resources and education levels needed to develop particular tools and knowledge (Chapters 6 and 7).

The OVER criterion of science establishes when we reach the boundaries of our scientific theories. The criterion can be applied to assess any scientific theory in any scientific field and research programme that aims to explain real phenomena in the world—such as physics, chemistry, biology, medicine, earth sciences, economics, psychology and agriculture, including the empirical and theoretical branches of these fields. It thus refers to any theory in empirical and theoretical physics, empirical and theoretical chemistry etc. It refers to *real phenomena we study and the theories of the world we develop about them*, and whether they represent observable, verifiable and reliable knowledge. It does not refer to abstract methods we create to study such phenomena. These include mathematics and logic that are not observable in an immediate sense and are not real phenomena in the world independent of us—in any sense comparable to scientific phenomena we study in the real world. All sciences—from natural to social sciences—reflect empirical or theoretical fields or both, whereas fields like mathematics, logic and computer science (at times labelled

Table 19.1 The present limits of science: three types of theories about phenomena in the world.

Type of theories about phenomena	Examples	Accessibility to phenomena	The OVER criterion of science: Phenomena are observable (directly or indirectly) at present	Theories (about phenomena) are verifiable at present	Methods and evidence (on which theories are based) are empirically reliable at present	
Direct observables (*Direct OVER phenomena*)	Plants, animals and habitats	Using our *universal methodological toolbox* (cognition and senses) to observe, experiment and solve problems	Yes	Yes	Yes	
Indirect observables (*Indirect OVER phenomena*)	Molecules, chemical compounds and viruses Mental states and economic markets	Using our *adaptive methodological toolbox* (methods and tools) including microscopes and statistics that we develop using our mind and that reduce cognitive and sensory constraints	Yes	Yes	Yes	
Unverifiables (*Non-OVER phenomena*)	The size and nature of the universe, dark matter and general causes of democratisation	Not currently accessible using cognition, senses or methods, but with direct evidence possible	No	No	No	← Present limits of science
	The historical origin of life, the moon and our evolved language, the evolution of consciousness, and the future evolution of the planet	Not currently accessible using cognition, senses or methods, and additionally constrained given spatio-temporal data collection limitations. That is, no direct evidence is possible, as phenomena lie in the past or future				
	Pseudoscientific phenomena such as homeopathy, astrology and classic psychoanalysis	Phenomena have been empirically tested using scientific methods and proven not to be reliable				

formal science) are especially used as formal methods or language tools that we create and, in contrast, apply within each of those fields of science. And our focus here has been on scientific theories that aim to explain real phenomena in the world—not on such methods we use to develop or evaluate those scientific theories or on other aspects of science.

Theoretical science accounts for a small share of all publications across scientific fields, as illustrated via the Scopus database (the largest database of scientific journals).[66] Publications on 'applied/empirical/experimental physics' (which do not include the terms 'theoretical/formal physics' in the publications) account for 94% of publications, with 6% of publications on 'theoretical/formal physics' (which do not include the terms 'applied/empirical/experimental physics' in the publications). This share of applied, empirical or experimental research accounts for 98% of publications in psychology, 97% in economics, 95% in medicine and 88% in biology (using the identical search terms with Scopus's default search function). Especially in theoretical science we are more likely to hit the limits of our theories and it is here where they are often most disputed, such as string theory. It is when theories involve phenomena that do not fulfil the OVER criterion. This criterion enables us to assess the reliability of all scientific theories across empirical science (that makes up most of science) and theoretical science (that make up a small share of science). Any researcher can apply the OVER criterion to assess any scientific theory that aims to explain a phenomenon across the physical, biological, behavioural and social sciences. Science is a pragmatic enterprise and the OVER criterion can thus be applied to likely over 99% of scientific theories, which is often as good as we can get in scientific practice. The small share of mathematicians, theoretical computer scientists and logicians who may not view their fields as methodological fields (which involve methods and tools applied across the natural and social sciences) do not need to apply the criterion.

We need to also view these features of science together. This is because when phenomena are not directly or indirectly *observable*, the theories we formulate about them are not *verifiable*, and the methods and evidence they are based on are not *empirically reliable*. How these features interact is through the scientific process of observing phenomena (or attempting to), formulating and verifying theories about them (or attempting to) and establishing whether the methods and evidence on which they are founded are reliable and replicable. This allows us to make predictions about theories. When we cannot fulfil the OVER criterion of science, we reach our present scientific limits. Ultimately, if phenomena in the world (P) are not directly or indirectly observable (O) via our universal or adaptive methodological toolbox (M), then we lack sufficient evidence (E) to develop theories about them (T) that are verifiable (V) and empirically reliable (R), which is when we reach the present limits of science (LS). Theories in science can be assessed using this limits-of-science rule:

$$\textit{If P is not O via M} \Rightarrow \textit{insufficient E for T that is not V and R} = \textit{LS}.$$

There is also a distinction we need to make among unverifiable theories. It is namely between cases in which we do not know how to develop the right theory about some difficult-to-comprehend phenomenon and cases in which we may have developed the right theory but we are not certain. In both cases the challenge is that we are not able to gather sufficient empirical data and verify them using our cognitive and methodological resources.

We are not directly able to perceive phenomena like DNA and quanta, but we eventually developed instruments and carried out experiments that enabled collecting sufficient empirical data to be able to test and verify them. Yet DNA and quantum phenomena were beyond the margins of science before we could collect data to confirm them. These are the distinguishing features of theories, namely: between those about presently non-observable phenomena that are non-testable and non-verifiable, and those that may become testable and verifiable as we develop new methods and experiments. The point is that these distinguishing features draw the defining lines between theories that are not reliable and those that may become reliable. In the case of DNA and quanta, we resketched the edges of the biological and physical sciences in the 20th century. We resketched the edges by developing new methods and instruments that enabled us to gather evidence about such phenomena that moved from being unobservable and unverifiable to indirectly observable and verifiable. For theories, or more precisely hypotheses, like those about superstrings and multiple universes, we do not (yet) have methods and instruments to empirically test them and they are thus outside the periphery of reliable science.

Yet given our remarkable methodological abilities of the mind, where does the mind end and the methods we create begin? We use our mind's internal abilities to develop complex external methods, and must interact with our natural and social environment to do so. Methods are, once created, external material artefacts in the world that can be shared and used by others—such as statistical and computational programmes. Sophisticated methods we develop facilitate the scientific process, such as generating hypotheses (using machine learning methods), collecting and cleaning data (using database systems), analysing data and simulating experiments (using statistical programmes). Yet we always use our mind's methodological abilities to develop such programmes and evaluate and interpret results, draw conclusions and assess ethical and policy implications. All research and discoveries in contemporary science require using both our universal and adaptive methodological toolbox.

While direct empirical evidence for phenomena is often lacking and abstraction can at times improve understanding about them, we eventually, when phenomena are less and less observable, arrive at the bounds of science. Ultimately, the foundation of science is thus empirical: if we cannot empirically test a scientific theory about real phenomena, if it is unprovable, then it cannot get beyond the realm of abstraction. While we increasingly require abstraction and imagination to study indirect observables, we often rely on high levels of abstraction and imagination when dealing with unverifiables. Yet in science, using abstraction and imagination to study the unobservable, when empirical testing is not possible, often remains largely that—abstraction and imagination. Our strongest theories about the world are founded on vast empirical evidence and reinforced using a variety of methods and data sources collected over time.

Consider a hypothetical scenario of science in which humans are conceived as the pinnacle of a meticulous designer's creation, rather than the result of an evolutionary process within our ecological and cultural context. Our visual capacity would be flawless, allowing us to perceive phenomena at the atomic level, the vast universe and beyond, rendering microscopes, telescopes and similar aids redundant. Our memory would be infallible, retaining precise recollections of all observations without reliance on datasets. Our processing capacity would enable rapid computation of thousands or millions of observations, making obsolete the need for computational tools and statistical methods to analyse complex phenomena. We would be endowed with an omniscient understanding of reality that surpasses the limitations of our human context. Such flawless abilities would facilitate unhindered scientific research and making flawless predictions about the future. However, the reality is that we humans have adapted evolutionarily to our ecological and cultural context, which has inevitably shaped our cognitive capacities, perception and the methodologies we devise to comprehend the world. Mind–method synergies set the present limits of science by shaping the methods that our mind is presently capable of developing to study the world.

Science's central evaluation criteria of testability, verifiability and reliability are grounded in our evolved mind and methods

How reliable our evidence and conclusions are is a question that all scientists must constantly deal with. This account of science here also explains how our evolved cognitive system lays the foundation for science's central evaluation criteria. Whether we view evidence or a theory about the world scientifically reliable depends on whether we, using our mind, are able to perceive the phenomenon, test and verify that evidence or theory and develop methods and instruments that are commonly needed to do so. Only with our human mind, methods and instruments can we do this. Science's central evaluation criteria of empirical testability, verifiability and reliability are embedded in our cognitive and methodological architecture.

Our cognitive and methodological limitations thus provide an empirically grounded explanation for how we evaluate science. It is an evolutionarily grounded and seemingly natural explanation. And the central features are captured in the OVER criterion of science that takes us further than other existing criteria of science, which we discuss below. When we study the extremely miniscule or large, the extraordinarily complex, or the very distant past or future, theories about phenomena are eventually no longer verifiable and reliable (Figure 18.2).

It is the theories of research programmes that move from being unestablished to established science, depending on whether we can eventually empirically verify them or not—and not just a property or feature but the entire theory. The reliability and accuracy of our theories depend on how well they fare empirically, with evidence in the world. Some are more reliable and accurate than others. We increase the reliability of our evidence and theories when they are consistent with one or more established areas of evidence and theories and when they are consistent using other methods.

This account places our methodological toolbox at the foundation of science and its evaluation: if theories about phenomena are not empirically testable and verifiable using our methodological toolbox, then they cannot be reliable—and thus cannot be consistent and replicable using different methods and evidence across different contexts. Yet why can't we just say that 'string theory is elegant mathematical formalism and does not yield testable predictions' or that 'we currently have no good theories of dark matter'? Why do we need the OVER criteria to support these widely held views among scientists? The OVER criterion provides a common measure to explain and justify why such theories are not yet reliable and it helps us settle debates in science over which evaluation criteria are necessary.

Defining the boundaries of science: shifting how we demarcate science from what is and is not scientific, to what is and is not presently scientifically reliable

Distinguishing between *scientific* and *non-scientific theories* is, for particular cases especially at the research frontier, not feasible nor would it be meaningful. We need instead to *shift the debate to distinguish between reliable and presently non-reliable theories.* This reflects a pragmatic shift from thinking about science in general idealised terms to doing so in practical measurable terms. This cognitive and methodological explanation here of the foundations and limits of science provides a new perspective to the demarcation debate—the debate about the boundary between the scientific and non-scientific. This question of how to demarcate science is one of the central long-standing debates in philosophy of science. Historians, philosophers and sociologists of science have attempted to address this question by arguing that what is scientific and what is not depends on whether it is or is not experimentation-based,(214) logically verifiable,(107) falsifiable (refutable),(15,33) puzzle-solving and hypothesis-revising,(208) guided by certain epistemic norms,(130) advancing a research programme(16) or reflected in systematic knowledge(269)—or they have argued that no criterion is needed.(17) The proposed criteria are narrow, each reflecting a single criterion, and they are also commonly based on theoretical and conceptual arguments—not empirical evidence.(31,259)

The present account is the first to be grounded in empirical evidence on the limitations of human cognitive, methodological and instrumental capacities. It provides a more empirical, evidence-based and scientific foundation for distinguishing between what is and is not presently scientifically reliable.(ibid.) This methods-driven perspective here provides a new way to demarcate the boundaries of science by placing the central role of our sophisticated scientific methods, instruments and methodological abilities at the centre of focus. We place here the powerful role of our methodological abilities and sophisticated tools and their constraints and potentials at the centre of defining what science is, its present limits and how to push those limits. This shifts the focus from such theoretical definitions to an active process of understanding and fostering methodological innovation to push our scientific borders. We connect here our mind's methodological abilities and sophisticated tools with empirical exploration and verification of phenomena they enable into a unified framework.

The central argument here is that we can best distinguish what is and is not presently scientifically reliable by evaluating when our theories involve phenomena that are not yet (directly or indirectly) observable and thus the theories are not verifiable and empirically reliable using our mind, methods and instruments. This OVER criterion of science is always needed for getting science off the ground—to be able to do science. It is as close as we may get to a universal definition or unified scientific methodology for evaluating theories that aim to explain real phenomena in the world. This criterion clarifies the distinction of the borders of science for old theories that have led researchers astray (such as geocentric theories of the universe, theories of spontaneous generation and the steady-state theory of the universe), for pseudoscientific theories (such as on homeopathy, astrology and classic psychoanalysis) and for particular historical theories in fields like biology, geology and cosmology for which we lack sufficient empirical evidence in the deep past. This criterion helps distinguish between reliable and not presently reliable theories in over 99% of cases in science. We may have to abstain from labelling particular theories developed by some research programmes at the research frontier as reliable knowledge until they may become empirically testable and verifiable. Ultimately, the OVER criterion helps distinguish between established science and those research programmes that have not yet been empirically established (that are at the research frontier) or that may not become established (that are beyond the scientific periphery). We shift the focus here from whether theories can in principle be tested to whether they can be tested in practice with our current or foreseeable methods and technologies.

In sum, the bounds of science, and what science itself is, can be better explained and understood in light of this account of what we are able to observe, test and verify, and what we are presently not able to, by applying our evolved mind and tools. The foundations and boundaries of our mind and methods thus shape the foundations and boundaries of our knowledge. We are only able to acquire knowledge about those phenomena in the natural and social world that fall into our present sensory, cognitive and methodological range that we are able to perceive and access. Where our mind, methods and instruments provide us evidence of the world, we can have scientific knowledge. Where they have not yet provided us evidence, we do not yet have scientific knowledge. *The history of science is a history about all the ways we have transcended the limits imposed on us by evolution and by our methods and instruments, at any given point, that we developed thus far.* This account of science is thus grounded in a seemingly natural explanation for both the foundations and limits of our scientific theories and thus science—namely the OVER criterion of science. By providing a new foundation for evaluating our scientific theories and science that is grounded in our evolution, history and empirical evidence, this criterion of science can resolve ongoing debates about influential theories. This includes debates about scientific theories that do not (yet) fulfil the criterion (such as on multiple universes and superstrings) and that have been proven not to fulfil the criterion (such as on homeopathy and classic psychoanalysis).

Most importantly, gaining deeper insight into the foundations and boundaries of science facilitates expanding science by directing our attention to our particular cognitive, sensory and methodological constraints and establishing ways to mitigate them. Each scientific method and tool, like our senses, has a given range and limitations, and we can extend that range by understanding those limitations and developing methods and tools designed to tackle them. We thus push the limits of science by creating new tools that expand our scope to the world.

20

The Limits of Science: Expanding the Limits by Expanding Our Methodological Toolbox

In every era, there exist scientists who believe that we have already largely solved the grand enigmas of the universe. Isaac Newton elucidated the laws of motion and gravitational force, while James Clerk Maxwell delineated the properties of electricity, magnetism and light. Dmitri Mendeleyev assembled the fundamental components of the periodic table that form the cornerstone of chemistry. Charles Darwin and Alfred Wallace outlined the evolutionary principle underpinning biology. Adam Smith explained the principles of the Wealth of Nations and the 'invisible hand' of capitalism: division of labour, freedom of trade and pursuit of personal interest driving societal benefit. These contributions laid the foundations for physics, chemistry, biology and economics, and many then regarded these foundations as essentially complete.

As the 19th century closed, the laws governing motion, gravity, electricity and magnetism, as well as conservation of energy and thermodynamics seemed to explain our entire physical reality. Physics seemed to have unravelled the fundamental laws that govern the world. In 1900, Lord Kelvin, a pre-eminent figure in the field of physics, asserted during a meeting at the British Association of Science: 'There is nothing new to be discovered in physics now. All that remains is more and more precise measurement.' However, the early 1900s witnessed the emergence of revolutionary theories such as relativity and quantum mechanics, which unveiled significant gaps in our understanding of our physical universe at both macro and micro scales. Einstein's relativity theory in 1905 introduced a novel way to perceive time and space, and in combination with quantum theory developed in subsequent years they revolutionised the field of physics. These advancements posed a challenge: Can quantum theory, which concerns the behaviour of particles at the subatomic level, be reconciled with relativity theory, which concerns the behaviour of massive objects on cosmic scales? The unification of electromagnetism with gravity remains an unresolved puzzle. That no major

Science of Science. Alexander Krauss, Oxford University Press. © Alexander Krauss (2024).
DOI: 10.1093/9780198937401.003.0021

breakthroughs in science remain to be made is a view held by many scientists at any point in history.

In this chapter we will describe how we can make advances at the scientific frontier quicker by describing the steps to extend our scientific tools to study the world in novel ways. We will also discuss whether there are pre-established boundaries to some domains of knowledge, and we outline pathways of the future of science that we can take.

Extending the present borders of science by extending our methods and tools

We continually make new scientific advances through our new methodological and technological innovations, and the pace of making such innovations has not decreased. In the past few decades, physics has been continually expanded. In 2002 a supermassive compact object at the centre of our galaxy was discovered by applying a new high-powered spectroscope, which revamped how we understand our galaxy and our place in it.[(270)] In 2015 gravitational waves were detected by applying a newly refined laser interferometer—100 years after Einstein developed his general theory of relativity in 1915 that predicted gravitational waves.[(246)] Biology and chemistry have also been continually broadened. In 2000 the mapping of ribosomes was completed by applying a newly upgraded electron density map, which fosters producing antibiotics to combat the bacterial infections we face. In 2012 the gene editing method CRISPR was discovered by applying a new differential RNA sequencing technique, which reshaped the life sciences, enables us to alter the DNA of plants and animals and facilitates the fight of cancer.[(245)] These major breakthroughs each earned a Nobel Prize and were brought about by inventing these new methods and instruments. A general explanation does not however exist for how we extend our scientific periphery which holds for all new major scientific advances.[(220,247,248,249,250,251,252,253,255,259)]

Here we provide a general explanation of how developing new methods and instruments is the central driving force of new scientific advances by enabling us to study the world in novel and innovative ways. We have observed earlier how science can be shaped by different factors (Chapters 2–15). Yet while our mind, methods, scientific community and broader economic and historical context influence science, the factor we can most directly influence is developing and applying new methodological innovations that enable making scientific advances. Newly created methods and tools are always

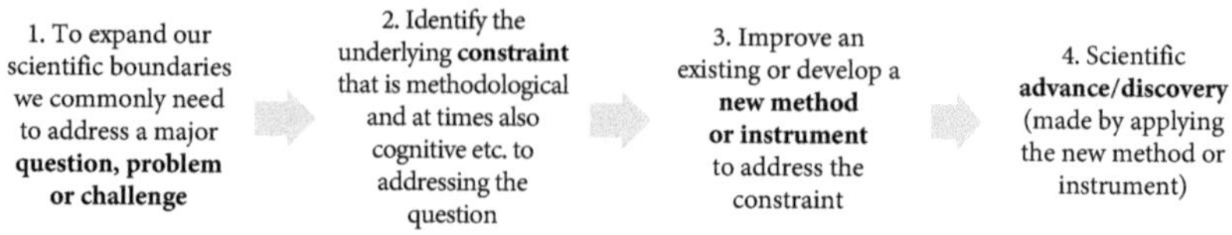

Figure 20.1 General steps commonly taken to expand the frontiers of science.

applied and are the factor we commonly require investing the greatest effort into (the time dedicated to developing and applying them).

Science's most powerful tools—from electron microscopes and X-ray methods to electrophoresis—were each developed in one field (such as chemistry) but later used to make discoveries within different scientific fields (such as biology, physics and medicine). These discoveries were not anticipated by the inventors of these powerful tools and were not the purpose of developing them. This highlights the direct causal link of new tools triggering scientific advances that they were not made for.

While developing new major tools generally updates our understanding about a part of the world, scientists generally concentrate on doing research and not on studying and conceiving how we expand our tools and the needed steps to do so. Yet this is required for us to expand our means of studying and understanding the world. Designing new methodologies and instruments that are better than existing ones can however often be challenging. As new methodological innovations drive new advances at the frontier, we need to address a critical question: What drives new methodological innovations and how do we make them?

Pushing and redefining the limits of science is achieved through major scientific advances—not by conducting conventional research that applies existing methods to create knowledge. To stretch our existing boundaries in a field and address a major question, we commonly have to identify and tackle the constraints to studying the question. Constraints are methodological and can simultaneously be cognitive, social, economic and related to other human features. We need to thus improve an existing method or design a new method to address the constraint, often iteratively through trial and error. We push science through this methodological process (Figure 20.1). The discovery process can at times begin with just this mechanism: breakthroughs are at times brought about by a new method or instrument that we use in an exploratory way, with no question being tested. Awareness of this mechanism enables us to take steps to address our constraints and extend our scientific reach.

Think of the spectrometer for instance. Kirchhoff and Bunsen created this instrument in 1859 which allows us to analyse wavelengths of electromagnetic radiation that we were otherwise constrained in perceiving. This powerful tool stretched our ability to investigate the structures of atoms and molecules and the chemical composition of planets and stars, pushing back the borders of the scientific periphery. Kirchhoff was thus labelled 'as Bunsen's greatest discovery.'(271) Yet the limitations of current spectrometers have become evident over time, including relatively low sensitivity,(272) often requiring combining them with other methods like chromatography to better probe more complex samples, and also computational constraints.(273) The spectrometer is just one example among leading scientific methods and instruments, and we discuss others later.

In general, when designing any scientific tool, we are asking how we can answer a question by improving our existing cognitive or methodological abilities to study the world. Establishing the methodological and human constraints we face is essential, as we cannot otherwise design a scientific tool that can mitigate a constraint that we are not yet fully aware of. We need to thus first generally detect a problem or limitation in the way we perceive or process phenomena in the world that we cannot currently explain, and then identify the basis of that problem or limitation.

The best researchers at the frontier are generally those who can best describe the gaps and complexities in our knowledge we have not yet figured out and then design and experiment with new methods and instruments that can fill those gaps, or collaborate with those able to do so. Making breakthroughs is about generating the right new methods and instruments that enable redrawing the lines at the edge of science.

This central point can be demonstrated with for instance the method of RCTs. Prior to conceiving RCTs in 1948 our ability to explain the causes of many medical diseases was limited. The RCT method—as most scientific methods—is the result of continual refinement over time of previously used methods and techniques. RCTs are, as outlined, an experimental method in which people are distributed randomly into treatment and control groups to test if a treatment or intervention may be effective. The creation of RCTs introduced innovative methodological techniques such as *randomisation*, aimed at mitigating confounding variables by randomly assigning experimental groups within a sample to ensure balance across those variables between the groups. Techniques of *blinding* were introduced to counteract biases like observer effects and confirmation biases by withholding information that can inadvertently influence the behaviour of participants and researchers in the study.

Placebos were also introduced to address the issue of participant expectations of a treatment impacting experiments by providing a group of participants with a placebo resembling the treatment but lacking a therapeutic effect.[42] Through iterative refinement and integration of these techniques into the RCT design, we gradually developed this composite methodology over time.

The RCT method has become the leading method across the biomedical, behavioural and social sciences for assessing how effective treatments and interventions are and the method has redefined the limits of knowledge in these fields. In this sense, the experimental setup of RCTs serves as a powerful success story of our ability to alleviate constraints and deliver life-saving evidence. And we will continue developing techniques to reduce the human constraints and biases we face.[42] The invention of randomised controlled experimentation exemplifies well how we devise methods and tools that broaden the horizons of scientific research.

The invention of novel methodologies and instruments commonly requires a point of reference rooted in existing tools. In our quest to expand comprehension across domains such as the complexities of the mind, the intricacies of life or the vastness of the universe, we require methodological scaffolding. This entails a process of exploration and experimentation aimed at innovatively building on and reconceiving our available tools to better explain a given phenomenon. The transition for instance from rudimentary trials in early medicine to the systematic evaluation of medical treatments through sophisticated controlled trials exemplifies the culmination of successive methodological enhancements and breakthroughs over time. These are closely linked to developing the RCT method. The same applies to the technological advance from horses to automobiles as a means of transportation, with our initial motor-driven vehicles bearing much resemblance to horse carriages.

Screening major discovery-making papers we can establish the ways new methods and instruments have been developed that sparked scientific advances. We observe that improving existing and developing new tools generally involves recombining and extending features of existing tools. We have extended the frontiers of science by expanding our methodological scope to the world through different pathways:

- employing methods including tools from near and distant scientific fields;
- integrating methods within and across scientific fields;
- experimenting with ways to upgrade our methods to address new questions and challenges;

- recognising the assumptions and limitations inherent in our methods and devising strategies to mitigate them;
- inventing completely novel methods for tackling new kinds of challenges we face;
- establishing our evolved cognitive, sensory and social constraints, and enhancing our methods to mitigate them;
- establishing strategies to address features of our spatio-temporal constraints.

A comprehensive explanation and framework has eluded us, until now, for how to expand our scientific frontiers by expanding our scientific tools.[220,247,248,249,250,251,252,253,255,259] We need to take such steps to broaden our powerful methodological toolbox. These pathways hold enormous potential to understand parts of the world currently beyond our grasp by offering novel perspectives. Much cutting-edge science entails experimenting with and using available methodologies from across disciplines and devising, probing and refining innovative methodologies, while applying much trial and error.

The question of how to maximise our cognitive and sensory capacities can appear strange, but addressing this question is essential to mitigate our cognitive and sensory limitations by developing tools that offer novel perspectives and processing capabilities. Identifying which constraints remain unaddressed is paramount because it opens uncharted research territories and enables making advances by strategically examining and establishing those constraints and experimenting with methods to mitigate them. Improved methods enable studying greater complexity, deeper into the micro or macro levels, further into the past, and so forth. Making advances requires continually reducing our methodological constraints. We have not yet been able to establish the best methodological means needed to address a given problem or its underlying constraint for many phenomena. Consider big data methods or network analysis methods used to study highly complex phenomena, which can involve as much noise and constraints as they hope to reduce. Tackling the existing limitations of computational power, machine learning, big data methods, network analysis methods and the like can play a significant role in fostering scientific progress across disciplines.

Next, we outline common methods and instruments of science and the main constraints they decrease, as well as their remaining constraints that need to be addressed to further expand science (Table 20.1). Making major scientific advances involves always reducing methodological constraints and at times simultaneously other constraints (cognitive, social, economic etc.). Shifting greater attention to such constraints and reducing them will enable us to access new phenomena and drive new advances.

Thus far we have outlined the scope and boundaries of science and can now better delineate what they mean. How we expand the frontiers of science is defined here as follows:

> Pushing the present limits of science is a cycle in which we have fewer methods and instruments available as we get closer to the boundaries of science, such as the electron microscope and statistical simulations. Here we have to identify which scientific methods and instruments we need to improve or newly develop to reduce our present cognitive, sensory and methodological constraints to studying and accessing parts of the world. And once we develop the new method or instrument—by building on and extending existing methods and instruments—we can access, measure and understand the given phenomenon in the world in new ways. We thus push back the boundaries of science, and the cycle begins again in an iterative process in which we continually expand science.

When researchers increasingly depend on abstraction and imagination such as in parts of theoretical physics and theoretical economics, they are more likely to hit against the limits of their field in the absence of empirical evidence. As the rate at which we generate complex computational, statistical and technological means is expanding quickly, science has the potential to continue expanding quickly. What are novel ways we can create scientific methods? We can for example increasingly carry out some studies using large-scale online collaborations with researchers linked via computers across the world and we can continually update such studies in real time (similar to the online encyclopaedia Wikipedia) and without a predetermined result. Such studies can provide a highly collaborative and iterative process in which replication and peer review are integrated in the research design. We can also increasingly use automated computational programmes to reduce some human biases in the scientific process. This includes routine processes that take much time and applies across all fields. This allows directing greater attention to creative aspects of the scientific process.

About everyday knowledge, Ludwig Wittgenstein stated that 'the limits of my language are the limits of my world.' About science and discovery, it is mainly the limits of our scientific methods and instruments that are the limits of our world. For our scientific tools are the present boundaries of how we are able to measure and analyse our world. We hit the limits of science as we hit the

Table 20.1 Creating new scientific tools advances science by mitigating our constraints to studying the world (examples of common methods and instruments).

Method or instrument developed to reduce constraint	**Year first developed**	**Constraints (methodological, cognitive, social etc.) that the method or instrument reduces**	**Phenomena that the method or instrument enables us to study or explain** (i.e. without which we would not be able to)	**Current main constraints of method or instrument** (that we must overcome to, once again, push the present limits of science)
Modern microscope	1873	Our limited visual capability to magnify and perceive very small objects	Observing microscopic phenomena such as cells, viruses and minerals; biomolecular interactions	Light microscopes have an ultimate resolution of ~250 nm, related to the diffraction-limited resolution etc. [274]
Computer	1940s/ 1950s	Our limited cognitive storing and processing capacity; mechanical processing of information etc.	Analysing large-scale data of phenomena across most fields, complex computational analysis, such as on climate modelling, genome analysis and econometric modelling	Limited computational speed per unit of energy/at fixed power supply; limited storage per unit of space; computation quantum speed limit etc. [275]
RCTs	1948	Our limited ability to control, randomise and blind participants to study them in a controlled experiment; biases such as observer effects and confirmation effects	Estimating how effective medical treatments and public policies are; more robust causal estimates in studies	Bias can arise due to sample bias (as some people refuse to participate), selection bias (as some are partially blinded/unblinded) and measurement bias (as some are treated for different durations) etc. [42]
Machine learning	1957	Human error and bias; influence of individual objectives and interests; limited processing capacity	Machine learning algorithms for rapid data cleaning, statistical analysis and identifying patterns in datasets without human influence; used in medical practice, speech recognition and data mining of population-level data	Automation bias; generally requires obtaining very large datasets; ethical considerations when applied to medical issues etc. [276]

limits of our best telescopes, particle detectors and statistical methods. It may thus appear obvious but an important insight that has not been given much attention in science is the following: the scope and limits of our present scientific tools are the extent we currently reach in explaining our scientific reality; they largely define the present scope and limits of science. Developing more sophisticated tools has enabled us to continually push back the present boundaries in all fields that would have otherwise been unimaginable (Figure 18.2). We cannot say much more about the scientific world than what our best tools enable us to say. The central message here is: *the key to expanding the limits of science is designing and experimenting with better methods and instruments—it is extending our methodological scope to the world.*

Do we face pre-established boundaries in science?

Our species faces inherent limitations in perception and information processing and has evolved to meet basic needs and reproduce, not specifically to understand the complexities of life, the mind and our physical reality. Does this imply predetermined boundaries to knowledge and science? Devising novel methods and instruments enables us, method by method, to extend the horizons of science. We go beyond our current methodological grasp by combining, expanding and devising improved methods. While our biological evolution, our cognitive limitations, our economic constraints and our spatio-temporal context delineate the realm of the world accessible to us, they do not dictate the kinds of tools we can conceive to mitigate those constraints and better comprehend the world.

The universe is finite. Life is not eternal. The speed of light is not infinite. The number of existing species of plants and animals is not limitless. The combinations of chemical elements are not boundless. The planet's natural resources we consume are limited. No organised complex phenomena, like the human mind or the universe, could exist or develop if there were no limits to what is possible in nature and if there was not much stability in the world. In our finite world, making major discoveries and opening scientific fields may thus also not be limitless and endless. However, our cognition seems less confined by explicit boundaries given our vast human creative imagination that we can tap to conduct science and devise new methodologies.

Generalising about the present limits of science can be difficult. Across different fields the complexity of phenomena we study and the presently available methods and instruments vary widely. Different fields of science face different data limitations. Are there for example limits to theoretical physics?

There are because our descriptions of phenomena are abstract and not testable when theorising about superstrings and multiple universes. Are there limits to palaeontology? There are since we can only investigate and reconstruct those past forms of life on earth for which fossils remain. Are there limits to cognitive science? There are given that we cannot recreate the origin of the mind or the evolution of consciousness, language and altruistic behaviour, and we study the mind from a self-referential point of view using our mind. Are there limits to cosmology? There are due to insufficient empirical data and a point of reference for events prior to the big bang. Are there limits to comparative biology? There are as we do not have systematic means of studying how other animals, with different sensory and cognitive abilities, perceive us and the world. There are limits to what we can predict across science, such as predicting well complex phenomena like ecosystems, economies and human behaviour.

Establishing major new fields that are as foundational as physics, chemistry and biology is also improbable—just like we are today not going to come across major new mountain ranges and rivers on our planet. Groundbreaking advances that generate a new basis for a major field are likely to decrease over time. Yet we will still generate new major breakthroughs and open smaller fields especially related to the range of phenomena on our planet. But much more is unknown beyond our planet and solar system. Is our universe finite or infinite? Do other intelligent beings exist beyond our solar system? Would different physical principles apply on different planets? Would different mechanisms of evolution exist for their species? Answers to such questions are currently beyond our methodological reach. For now, nearly all of science is based on a single dataset, our own planet, and its particular ecosystem, characteristics and evolutionary processes. This dataset basically accounts for our entire body of knowledge about the world. Coming across other species on other planets and galaxies and being able to conduct experiments there would be a big revelation for science—to be able to test, compare and update our best theories about the world. Science itself would no longer have just one experimental group but entirely new experimental and control groups (other planets and galaxies). Some knowledge we have acquired up to now would become more robust, if the evidence would be the same for physical phenomena and species on other planets, while other knowledge would need to be revised.

A paradox of science is that each advance brings us into deeper complexities that are presently beyond our reach. As our knowledge expands, generally so does our awareness of what we do not yet know. At the border of science, developing new advances and tackling problems generates new problems and puzzles. At least in some fields, the pace at which we push the research frontier will likely eventually become slower and slower. An indication of when a scientific field may partly near its present limits is when it becomes increasingly saturated with abstract theories that cannot be empirically verified. This appears to be the case in parts of theoretical physics (such as string theory, theories of multiple universes and dark matter) and theoretical economics (such as the dominance of rational choice theory). String theory is for example a long-sought attempt of a unified theory of physics. It aims to combine quantum mechanics with the general theory of relativity, in which particles are modelled as string-like entities rather than point-like particles. Yet as it is largely a mathematical construct and not verifiable with experimental evidence, it is more an idea than a testable theory.(277) String theory is a classic example of how we hit our present limits of science and how science breaks down, in the absence of empirical evidence. Another indication of when a scientific field may partly near its present limits is when major discoveries are not made for long periods of time despite many scientists and much research in the field.

Our attention in science will always remain on those aspects of the world that our mind and the tools we can devise with our mind enable us to access. Their boundaries are our current boundaries that we continuously reshift. Some questions and parts of the world—independent of the amount of researchers, methods and resources we direct towards them—may remain beyond our cognitive and methodological scope. These include for example what came before the origin of our universe and the size of the universe. Ultimately there will also be things we will not know. But we will not know that we do not know them. There are thus unknowns that we are confident exist but do not yet have the needed tools to uncover them—such as dark matter. There are imaginables that we can imagine but cannot yet test empirically—such as life beyond our planet. And there may also be unknowables that lie beyond both our methodological reach and imagination.

A methodological barrier in science is that it investigates phenomena that we can observe and are methodologically tractable and mostly quantifiable. Consequently, phenomena like the nature of time, knowledge, human consciousness and experience, freewill, politics or love pose considerable challenges for scientific investigation. In fields like philosophy, epistemology and ethics, what we study often lies beyond the scope of empirical evidence.

While our tools largely dictate the existing barriers of scientific disciplines, they are not able to elucidate all aspects at the scientific frontier. The barriers are also partly influenced by the allocation of policy attention and funding to particular fields and tools, especially relevant in subfields of astronomy and physics (Chapter 7). The barriers are also partly delineated by ethical considerations, including decisions of regulatory agencies on the legality of human cloning, gene editing, artificial intelligence, nuclear energy, military research and technologies impacting climate change. The boundaries of scientific exploration vary across different fields, with some more limited than others, and the extent to which we continue expanding those boundaries mainly depends on the ways we identify to extend our methodological scope. Ultimately only those researchers optimistic about the prospect of new breakthroughs are incentivised to explore and broaden our methodological horizons and scientific scope.

The future prospects of science: three pathways

We are confronted with more uncertainty the greater we look into the future, and we can either choose to not say anything about the future given the degree of uncertainty or we can discuss possible pathways of the future of science. Discussing such pathways can foster better planning the future of science. We opt here for the latter, which moves us into deeper philosophical terrain. We outline here three pathways of the future of science that we can take to different degrees in different fields:

- the filling-in-the-details view of future science;
- the empirically limitless but theoretically limited view of future science;
- the methodologically limitless view of future science.

The first possible pathway is the *filling-in-the-details view of future science* that ultimately, in possibly a few hundred years, nearly all major methods, instruments and discoveries would have been made and science would mainly involve smaller but important details and refining existing findings. The fundamental puzzles and grand breakthroughs would have been resolved—as currently with quantum mechanics, relativity, evolution, DNA, the periodic table and the big bang—leaving little room for us to make new foundational discoveries that entirely replace them. Debates on human evolution would for example reduce to debates about details of particular mechanisms of evolution. Measurements and results would however continue to be updated as we improve our methods and instruments.

The second possible pathway is the *empirically limitless but theoretically limited view of future science* that science would eventually focus largely on new empirical applications to changing environmental, medical, technological and social challenges of our time. All major theoretical discoveries would have eventually been made—as some already argue.[cf. (253)] Theoretical sciences would eventually decelerate, since establishing foundational theories, in fields such as physics and chemistry, makes establishing new theories less likely. Constantly creating grand theories that explain reality well and that constantly replace themselves would be inherently contradictory and is highly unlikely. The theory of evolution for example serves as the cornerstone of biology and is deeply embedded in independent strands of evidence using methods from molecular biology, palaeontology, primatology, cognitive science and archaeology. The periodic table of elements accounts for the elements that make up the world and thus the foundation of chemistry. This makes it highly unlikely that novel fundamental theories will completely replace the theory of evolution, periodic table of elements or other central theories of science. Just as we can only discover once our genetic code or that DNA has a double helix structure and transmits genetic information, we will not limitlessly make breakthroughs about the foundation of biology (evolution), chemistry (the periodic table) or physics (relativity and quantum mechanics). But they would remain similar to our current well-predictable theories, though become more detailed. Major new tools and breakthroughs have been and will continue to be developed among those grand theories and advances—such as CRISPR gene editing, mapping the human genome and the existence of the Higgs particle—that were made since 2000 and have earned Nobel Prizes.

Researchers across fields like medicine, biology, psychology and economics will likely explore new questions as long as our species exists. The enduring importance of such research is driven by the demand for novel treatments to combat emerging viruses, diseases and cancers, and changes in environmental and social conditions over time and context. Researchers in fields like agricultural science, environmental and mechanical engineering, computer science and technology-related fields will likely open new questions indefinitely, given their pivotal roles in enhancing human welfare and addressing our changing human, geographic and social challenges. Science and technology are highly interconnected. The perpetual quest for enhanced health, technological advancements and improved human well-being ensures continued growth in applied sciences. In the same way that the inventions of the internet, space shuttles and atomic energy were not foreseeable a century ago, many future scientific and technological developments remain equally unpredictable. Although some may believe we are reaching a plateau in scientific growth given that major theories are not likely to be abandoned,[(253)] the trajectory of applied sciences and technological progress shows no signs of slowing down in the future, and we observe continual advances. Embedded within this unified perspective of scientific and technological advancement lies the pivotal insight: it is often methodological and technological advances—such as enhanced telescopes, large-scale spectrometers, innovative machine learning methods and novel particle detectors—that have driven and continue to drive new major breakthroughs.

The third possible pathway is the *methodologically limitless view of future science* that we will continue to create currently unimaginable methods and instruments and integrate them in seemingly limitless ways across diverse domains. Generating new tools will enable us to continually expand existing and create new domains of science and discoveries given our endless imagination. It would seem at first glance contradictory for the world to be finite, but not our knowledge of the world. Yet we investigate the observable world by applying our flexible cognition and tools. The nature of knowledge is cognitive and methodological and thus variable, with scientists applying creativity and critical assessments that are not inherent in the world outside. Using human imagination, we can expand available tools to devise new tools that are seemingly unbounded by predetermined constraints in light of our seemingly boundless creativity. Thus even if the world is not boundless, science to some extent is, in light of our elastic cognition and tools.

The future of science will likely encompass, to different degrees in different fields, all three pathways that each fundamentally relies on the scientific methods and instruments we are able to develop. Elements of the three pathways will thus likely take place in all fields. The future of science depends on what we can methodologically imagine and create. Endless major theoretical innovations are unlikely in all three pathways. How we answer the question of whether we may be getting closer to the boundaries of science and what the future of science holds relies on the part of science we turn our attention to: whether we view scientific progress as more closely tied to new major empirical breakthroughs and the application of science, or to the creation of new major methods and tools that trigger breakthroughs, or to fundamental theoretical breakthroughs, or all of them. Theories are not however able to explain or justify themselves but always rest on empirical evidence and using tools and they generate questions that only empirical advances can tackle. Novel questions and tools we create generate in turn further questions and tools and opportunities for novel knowledge. Science will be around as long as our species is around, since our societal needs change and new challenges emerge. We always have novel medical and technological problems to solve. Much of science will not ever be finished. Overall, these are different pathways for how we can develop our tower of science in the future—that is, the pathways for the future of science.

A question of history of science is whether our theories of the world will appear as incorrect to our descendants in a few hundred or thousand years as Aristotle's theories appear to us? Yet the difference is that Aristotle's theories of motion, four elements (earth, water, fire and air) and biological theories could not accurately explain and predict our physical, chemical and biological reality. But today's methods and theories enable us to do so with high precision, so differences in accuracy will likely be much smaller between the theories of today and the future. This is a mark of the great scientific progress we have made thus far.

Vast growth has characterised science over the past century. Yet is it possible for scientific progress to continue as rapidly in the future? After making a discovery, such as the electron, particle accelerator and DNA molecule, they can no longer be discovered again. Yet such major breakthroughs generate new disciplines that cumulatively grow over time. Numerous novel fields of science are continually arising, such as computational biology, environmental science, artificial intelligence/machine learning, quantum chemistry, nanotechnology, genetic engineering and science of science. Such fields hold promise for novel scientific breakthroughs beyond our current imagination, by integrating, specialising and adapting to new conditions in society and our environment. At the forefront of science, progress appears promising. On the one hand, the prospect appears positive because the rate at which new pivotal breakthroughs and fields have emerged has not shown signs of decelerating in recent decades. On the other hand, the systematic development and dissemination of new methodological advances has yet to be carried out in a coordinated and targeted way across science despite their pivotal role in driving scientific advances. Awareness among the scientific community about this methods-driven nature of science has been lacking. The growth of science will be intricately tied to the continual expansion of our methodological toolbox.

In sum, we humans have evolved astonishing methodological abilities that enable us to do science and understand that we and our mind are the result of an evolutionary process and live on a planet revolving around a star within one among millions of galaxies in a vast and expanding universe. We have not yet fully addressed foundational questions about the nature of life, matter, the universe and science, but we do know a lot about them. Science possesses an inherent capacity for self-renewal; science will not culminate in a completed endeavour. We have achieved the majority of our major breakthroughs across science only since the 20th century, and the future is much longer and holds many more scientific advancements through powerful new tools we devise. The horizon of scientific exploration and future knowledge knows no definitive boundaries, and we have outlined here three pathways of the future of science. Ultimately, understanding everything about the universe would imply that we not only have answered all existing questions but also have established all possible questions and devised the needed methodologies and instruments to tackle them. The forefront of science is characterised by posing the right questions and conceiving the right tools—both small and large. Pursuing novel questions characterizes the common approach at the frontier, which is observed in scholarly publications, research initiatives and funding proposals. This common strategy, when attention is also directed to methods and instruments, can help us conceive and develop the methods and instruments needed to address such questions. An alternative strategy has been at least as successful, if not more: focusing our attention strategically on upgrading our existing tools and creating new ones that answer our existing questions and can answer completely new questions not yet raised. The central conclusion is that the best way we have to achieve new scientific advances is through the incredible power of developing new methods and instruments that have continually reset our present scientific boundaries. Elucidating here this methodological mechanism propelling scientific exploration puts us in a better position to achieve new breakthroughs and achieve them quicker. This central insight has not been given sufficient attention in studies in science of science or the limits of science to date.[220,247,248,249,250,251,252,253,255,259]

In general, we humans are what set the present scope and limits of science: the methods and instruments that we have been able to develop thus far, using our mind and within our broader social context. We ourselves are, through our methodological toolbox, the origin and boundaries of science—the origin and boundaries of what we can perceive, measure and understand in the world. Methodological advances allow us to greatly extend those boundaries by reducing our human and methodological constraints and broadening our scope to the world. Predicting the rate of new breakthroughs hinges on the rate of new methodological innovations.

This methodological mechanism of expanding our scientific frontier has vast ramifications: scientists need to focus greater attention and time to detecting our present methodological shortcomings to investigating the world and tackling those shortcomings to spark new advances. We have described here how we can push the current limits of science more rapidly and in a more structured way by outlining the steps we need to take to extend our tools of science. This shift in our focus needs to be combined with education and university systems also placing a greater focus on students better studying the methods and instruments used in science, their present limits and how to improve them. We need to train future scientists to better detect problems with our best tools of science and how to address those problems by experimenting with new methods, tools and techniques. Targeted public funding schemes for experimentation in and advances in methods and instruments are also needed to foster this shift across science. Strategically fostering this methods-driven mechanism of science could mark the beginning of a methodological revolution in science that changes the way we understand and do science and accelerates the way we advance science.

Conclusion and Implications

Viewing the field of science of science through an integrated lens can provide answers to fundamental questions about science: its origins, foundations and limits and how to push those limits. To date, these questions have commonly been studied by adopting a perspective from an individual discipline. Scientometricians, historians of science, psychologists of science and other researchers carrying the title '... of science' have thus not yet been able to provide a comprehensive and integrated account of science. The central challenge of the field of science of science has been accounting for and integrating the different empirical and theoretical knowledge across disciplines into a holistic field and uncovering the general mechanism driving science across fields. This book has aimed to tackle this challenge and offer a foundation for this integrated field by combining methods and evidence from across the natural, behavioural and social sciences (Figures 1.2 and 16.2). Adopting such a holistic approach has enabled us to file down the often inflated role of a single factor and assess which factors are most important and how they fit together, to then be able to develop a deeper and more coherent understanding of science (Figure 16.2). Science of science, conceived here as an integrated field, provides a unified understanding of science and how to improve science by identifying the abilities and conditions that drive and constrain science. The different subfields of science of science use different methods and study different aspects of science, and the evidence from the independent strands are coherent with what is already known across fields (Chapters 2–15).

The central conclusion is that the powerful role of our *universal and adaptive methodological toolbox* has been identified across fields as the main mechanism driving science that we can directly influence. Our methodological toolbox underlies the different factors across disciplinary perspectives. We observed that the central factors that have been proposed as the most important single factor explanation of science—namely the paradigm shifts that define scientific theories,[1] the principal evaluation criterion of science,[14,15,215] the key social influences on scientists[10,11] and so on—are not able to explain as much and do not have direct influence on the foundations, limits and advancement of science compared to other factors. These

Science of Science. Alexander Krauss, Oxford University Press. © Alexander Krauss (2024).
DOI: 10.1093/9780198937401.003.0022

factors proposed by the most cited researcher studying science within a particular field—namely Kuhn in history of science, Popper in philosophy of science, Latour and Woolgar in sociology of science and so on—need to be left in the background. Our methodological toolbox needs to be brought into the forefront of how we understand, study and advance science.

So what do the economist of science, scientometrician, philosopher of science and other individual researchers gain that they did not have before the integration of these fields into a holistic science of science? For some researchers it is a shift towards a joint research focus on methodology and better understanding and improving our best methods and instruments that drive science and discovery. For other researchers it is addressing the fractured approaches and filing down the rough disciplinary explanations about an often overemphasised role of power by sociologists, the truth of theories by philosophers, citations by scientometricians and so on to develop more integrated explanations. For all researchers it is a shift to studying science in an integrated way, coherent with already common knowledge in other subfields, rather than in isolation, whether they study meta-level questions about science or specialised questions. As making major discoveries has required us to apply new scientific methods and instruments, we cannot adopt an understanding of how we make discoveries that is independent of our scientific tools. But existing accounts of science (scientometric, historical etc.) have not yet incorporated the essential role of methods and tools into such accounts to date.

This *new-methods-drive-science* theory presented here illustrates how our mind and sensory abilities (to observe, experiment and process information) make doing science possible but also shape what and how we observe and reason, as they have evolved within our environmental niche. Our scientific methods and instruments (from new statistical methods to electron microscopes) enable us to study a much broader range of phenomena, but they also have constraints to how we measure and perceive phenomena and express our theories. Institutions, funding and societal challenges help influence what knowledge and research methods we produce, distribute and use. Scientific norms and methodological assumptions shape the way we evaluate our evidence, among other influences (Figure 16.2). Taken together, this theory explains how sophisticated methods and instruments we develop using our mind's methodological abilities set the scope and present limits of what we can know and what is possible in science—and economic, social and historical influences help shape what we study within that scope and those limits.

The origins, foundations and limits of science can be better explained, understood and advanced in light of the new-methods-drive-science theory. It illustrates that our scientific methods and instruments are the only factor that underpins all 14 subfields. With the foundation, boundaries and advancement of science largely shaped by our methods, a central focus on methods across all subfields is also crucial to the integration of the science of science. This methods-driven understanding of science can help spark a methodological revolution in science. We have thus seen how the *tower of science* we have reconstructed is made possible by our evolved methodological abilities (that account for its foundation) and our scientific methods and instruments (that account for the different floors of its structure). Together, they largely determine how far we can perceive, measure and understand the world. Overall, four main implications arise of this meta-approach to studying science.

1 *We need to measure the success of the field of science of science by establishing a society, journals, conferences and interdisciplinary institutes that adopt a truly integrated approach to studying science.*

These would help tackle constraints confronting the interdisciplinary nature of the field. They include methodological obstacles that involve researchers acquiring skills in multiple methods and collaborating in teams adopting multiple methods. They include institutional constraints that involve collaborations within cross-disciplinary institutes and greater access to established sources of research funding. While the status quo of isolated fields studying science has not led to integration, establishing a society, journals, conferences and institutes (or centres) in science of science would institutionalise collaboration networks and a forum of integrated debate. Establishing which disciplinary fields and factors are most important in helping advance science is key to the field (Table 16.1). It can enable scientists and scientific institutions to better direct their attention to particular features of science. Scientific institutions like the European Commission, National Science Foundation, governments and other funding agencies need to begin incentivising integrated meta-scientific research and new methodological research that is as important as other established funding areas in advancing science and discoveries. Leading scientific journals need to publish not only new scientific breakthroughs but also new methodological breakthroughs, as they are what drive new scientific breakthroughs—at times independent of pre-existing questions or hypotheses. We need to also revise the mainstream view of science and science policy that conceives scientific research as question driven.

Shifting much more attention from question- and hypothesis-driven science to methods-driven science that is exploratory can help foster new advances.

2 *We need to better train researchers studying science and better conduct research in a more interdisciplinary way.*

Basically all researchers studying science currently pursue disciplinary specialisation that can serve an important function, namely the division of cognitive labour by allowing researchers to concentrate deeply on a given topic. In many research domains this has been a very successful strategy—think of particle physics, molecular spectroscopy or brain surgery without extensive specialisation. But using this strategy to study a phenomenon as multidimensional and complex as science is the reason why we, to date, have not yet developed a coherent general theory of science and how science advances—the central question in science of science. There is no way around it: if we want to better understand science, we cannot resort to just saying that time and resources constrain us in working more interdisciplinarily. This has come at the cost of at times contradicting evidence across subfields on the same topics—that is, non-replicable evidence across fields.

We need to make a better effort and reform the way that science has been studied to date. The vision of science of science outlined here would ideally consist of perhaps roughly three-fourths of researchers studying science by continuing to pursue disciplinary specialisation but all researchers would begin to be trained in and spend a share of their time to be broadly informed about existing research on the same topic across other subfields of science of science. This is the only way to ensure a coherent understanding of science that is consistent, not conflicting across neighbouring subfields. And importantly, we would otherwise not even know that we have gained a coherent understanding of science, unless we compare the findings and evidence to assess if they are coherent across the subfields of science of science that use different methodologies to address the same question. The remaining perhaps roughly one-fourth of researchers would ideally then adopt a meta-perspective that truly integrates fields. This however requires that some science of science researchers would be formally educated and trained, through university departments and institutes, just as biologists or statisticians are trained—with degrees in science of science. We cannot assume that some researchers who happen to be inclined to adopt an integrated big-picture approach to science will come across such questions and find the time to dedicate themselves in a full-time capacity to addressing them. Just as there are general meta-researchers in environmental sciences, we also need to begin training such general meta-researchers in science of science, if we want greater depth and breadth in understanding science.

3 *By better understanding the set of constraints we face in science, we can reduce them to advance science.*

We can mitigate constraints and biases facing our cognitive and sensory abilities by making efforts to develop methods that expand these abilities. We can decrease constraints and biases facing our methods by improving them, applying multiple methods and developing new methods. We can reduce some social and contextual influences we face by conducting and comparing studies in different contexts and time periods. We can address psychological biases we face by automating some processes using computers and by different scientists replicating studies and thus confirming or refuting the results of others. We can mitigate our assumptions by expanding our methods, and so on. Essential to improving science is an awareness of the importance of developing new methods and instruments designed to reduce our human and methodological constraints to studying the world (Figure 16.2). With such an awareness, we can strategically target and address our constraints to advancing science. To push the limits of science, we need—as the central driving force—to extend existing methods and instruments and develop entirely new ones (Chapter 20).

4 *We need to revisit our best methods for assessing and measuring science and discoveries and adopt a broader set of empirical methods from neighbouring fields to better address foundational questions about science.*

To date, researchers studying science (including scientometricians, network scientists and psychologists of science) commonly do so descriptively studying one factor. Most studies in subfields of science of science, and nearly all studies in scientometrics, are descriptive studies that use observational data. Such observational and big data studies can uncover important and at times strong relationships among factors. But a critical step to move the field forward will be conducting studies that can assess causal effects of central factors driving science and their interrelationships, as in most human sciences.[4,5,35,39] Improving our understanding of science requires turning to methods that enable us to measure causal relationships better—such as those widely used in economics, public health and the medical sciences (outlined in Chapter 7).

We need to begin conducting for example longitudinal studies that follow scientists over their lifetime to assess long-term changes over their careers: including the effects of their discoveries, new collaborations and relocating to top universities.

To better understand the range of factors that shape science and its foundations and boundaries, we will also have to begin applying a wider range of methods from across subfields: longitudinal observational studies, instrumental variable methods, natural experiments, randomised controlled experiments, institutional analyses and so forth. Science of science is about applying science to understand science, and to do so better we have to move beyond descriptive big-data methods—as just one method among other methods. We cannot narrowly view science of science as the 'field that relies on big data to unveil the reproducible patterns that govern individual scientific careers and the workings of science' by studying primarily citations,[(4)cf.(5,35)] because big data can for example rarely provide causal knowledge that will require us to begin applying such experimental methods of medical statisticians and economists. For a number of big questions in science of science we will have to rely on observational studies and will not be able to apply experimental studies. But we will have to begin applying observational studies in a more integrated and interdisciplinary way.

Causal understanding is more difficult to attain in science of science than in other fields, given the nature of science as a complex phenomenon with multiple influencing factors at different levels. This makes it more difficult to compare how important different factors are against each other (such as newly developed experimental methods, cognitive biases, public funding, and team and community size). This is because they are captured at different levels using different types of data (such as experimental, individual, institutional and country-level data). Causal identification can thus be partly constrained when studying particular large-scale phenomena like scientific communities, the evolution of scientific fields or the efficiency of research institutions, or predicting the future of science. It can be constrained in addressing questions beyond well-defined, quantifiable interventions at the level of the individual scientist. The more complex and bigger the phenomenon we study, the greater the uncertainty we often have—think of networks of scientists and institutions, and novel breakthroughs in science. For such complex phenomena, we cannot generally apply randomisation techniques that are used to identify causal relationships, as we do not usually have a comparable counterfactual. Scientists cannot be easily randomised into treatment and control groups to assess for example the effect of making a major discovery or receiving a Nobel Prize on one's career or on science. For the sample would have to be near universal of the entire scientific community and thus include millions of scientists. Measuring science better requires that we better conduct studies assessing causation. There is no way around it: for science of science to reach its potential it will have to move beyond its strong focus on observational studies and conduct studies that assess the causal effects of interventions driving science. Measuring science better will also require us to go beyond the common use of citations as the key impact metric of science[4,5,39] that needs to be viewed as just one among other measures of scientific impact and advancement—such as scientific discoveries, methodological advances and social and policy impact.

Qualitative studies (that integrate insights from across disparate fields) are also important. However, constraints of such studies include quantitatively measuring factors studied across different fields (norms, institutions, practices, assumptions etc.); balancing more breadth across more fields with inevitably less depth on any single field; and synthesising across fields, as each field has its own methodological preferences and complexities. As highlighted up front, a companion book *The Motor of Scientific Discovery* assesses over 750 major scientific discoveries including all nobel-prize-winning discoveries and provides statistical evidence of how we drive science and discovery through new methods and instruments — and complements this book with such quantitative data and grounds this theoretical framework.[64]

More generally, improving science, the structure of science and fostering interdisciplinary science are important goals for advancing knowledge and addressing complex problems, whether in science of science or other fields across science. What are general steps we can take to improve and advance science, and also do so in interdisciplinary ways? One, we need to continually foster and adopt emerging technological advancements and digital tools that can enhance the way we conduct scientific research, data analysis and disseminate knowledge. These include high-performance computing, data analytics, imaging techniques and other cutting-edge technologies that can bridge disciplinary boundaries. Two, we need to better incentivise interdisciplinary collaboration and provide resources for researchers from diverse fields to better work together to address complex challenges. Fostering interdisciplinary research centres can play a key role in bringing together scientists from across fields and provide a platform for interdisciplinary collaboration, knowledge, resource sharing and problem solving. We need to support cross-disciplinary funding programmes, training opportunities, conferences and workshops, and reward interdisciplinary advances. We require also developing better educational programmes that expose students and researchers to interdisciplinary approaches early on. Yet interdisciplinary research can generally be more complex and require more time, effort and resources.[278,235,72] Three, we need to ensure research funding supports a broad portfolio of science projects, not just mainstream question-driven projects but also exploratory, higher-risk, interdisciplinary or longer-term projects that are less commonly funded. Four, we need to better support early-career researchers through career development opportunities and university mentorship programmes that connect young researchers with experienced researchers to provide guidance. Five, we require fostering efforts to improve science communication and its impact by bridging the gap between scientific studies and their use among policymakers and the general public. Promoting simple language in publications would increase public understanding and engagement.

These different measures require collective efforts from researchers, institutions, funders and policymakers to continually make our scientific ecosystem more robust and effective. Implementing these measures can enhance the scientific system and create an environment conducive for interdisciplinary science, whether in science of science or other scientific fields.

Ultimately, the integrated field of science of science holds a vast potential for addressing fundamental questions about the origins, foundations and limits of science. It holds a vast potential for better understanding how we drive new scientific advances and methods that open new and unknown frontiers.

Appendix

A brief clarification is provided here for how the share and number of publications across the subfields of science of science have been calculated (Figure 1.1). Publications, which include articles, books and other scientific formats, were identified by searching a given term in the title, abstract or keywords (Scopus's default search function). Data were collected for the fields that contribute to understanding the origins, foundations or limits of science; data reflect the outcome of searches in early 2024 using the terms 'biology of science' or 'philosophy of biology' (656 publications), 'cognitive science of science' or 'philosophy of cognition/mind' (3084), 'psychology of science' or 'philosophy of psychology' (456), 'linguistics of science' or 'language of science' (638), 'sociology of science' or 'philosophy of sociology' (1604), 'economics of science' or 'philosophy of economics' (319), '(scientific) methodology of science' or 'philosophy of (scientific) methodology' (300), 'computer science of science' or 'philosophy of computer science' (50), 'statistics/mathematics of science' or 'philosophy of statistics/mathematics' (1225), 'scientometrics' or 'network science of science' (5418), 'philosophy of science' or 'metaphysics/ontology/epistemology of science' (10,524), 'anthropology of science' or 'philosophy of anthropology' (137), 'history of science' (8268), 'archaeology of science' or 'philosophy of archaeology' (23), and 'science of science,' 'metascience' or 'metaresearch' (956).

Here we also briefly outline common methods used in each subfield (Figure 1.2). Biology studies science using methodological approaches such as comparative analysis of the abilities of human and non-human animals and an evolutionary theoretical framework. Cognitive science studies our cognitive and sensory abilities for perceiving and reasoning about reality. Psychology conducts experimental studies of reasoning, biases and motivation and uses direct observation in labs and surveys of scientists. Linguistics analyses the communication of science and language as a thinking tool in science and as a tool to accumulate knowledge. Sociology carries out quantitative and qualitative surveys of the scientific community, practice and norms and carries out institutional analysis. Economics conducts quantitative analysis of productivity, the reward system, research funding, science policy and institutional analysis. Methodology covers methodological analyses and studies the drivers (such as methods and instruments) of new discoveries. Computer science analyses computational methods (used to process and analyse data) and their foundations, including artificial intelligence. Statistics/mathematics analyses the foundations of statistical and mathematical methods and their constraints and biases. Scientometrics including network science conducts citation and publication analysis and uses large-scale data. Philosophy covers theoretical and normative analysis of science and its methods and analysis of assumptions, concepts, causation and laws. History carries out case studies of scientists and discoveries (autobiographies, notebooks) and historical analysis of scientific progress. Anthropology involves case studies and past and present cross-cultural studies (e.g. on reasoning). Archaeology analyses fossil records and material artefacts and tools (e.g. related to our ancestors' cognitive abilities).

References

1. Kuhn, Thomas. The Structure of Scientific Revolutions. Chicago: University of Chicago Press (1st edition 1962; 3rd edition 1996).
2. Daston, Lorraine, Peter Galison. Objectivity. New York: Zone Books (2007).
3. Daston, Lorraine, Michael Stolleis. Natural Law and Laws of Nature in Early Modern Europe: Jurisprudence, Theology, Moral and Natural Philosophy. Milton: Routledge (2017).
4. Wang, D., A. Barabási. The Science of Science. Cambridge: Cambridge University Press (2021).
5. Fortunato, S., C.T. Bergstrom, K. Börner, et al. Science of science. Science, 359 (2018).
6. Zeng, An, Zhesi Shen, Jianlin Zhou, et al. The science of science: from the perspective of complex systems. Physics Reports, 714–715, 1–73 (2017).
7. Sinatra, R., D. Wang, P. Deville, C. Song, A. Barabási. Quantifying the evolution of individual scientific impact. Science, 354, aaf5239 (2016).
8. Azoulay, P., J. Graff-Zivin, B. Uzzi, et al. Toward a more scientific science. Science, 361 (6408), 1194–1197 (2018).
9. Park, M., E. Leahey, R.J. Funk. Papers and patents are becoming less disruptive over time. Nature, 613, 138–144 (2023).
10. Latour, Bruno, Steve Woolgar. Laboratory Life: The Construction of Scientific Facts. Princeton: Princeton University Press (1979).
11. Bourdieu, Pierre. Science of Science and Reflexivity. Chicago: University of Chicago Press (2004).
12. Pickering, Andrew, Keith Guzik. The Mangle in Practice: Science, Society and Becoming. Durham, NC: Duke University Press (2009).
13. Collins, H.M. Scientific knowledge, sociology of. In: Wright, J.D. (ed.), International Encyclopedia of the Social and Behavioral Sciences. Amsterdam: Elsevier, 2nd ed., Vol. 21, pp. 308–312 (2015).
14. Popper, Karl. Normal science and its dangers. In: Imre Lakatos, Alan Musgrave (eds), Criticism and the Growth of Knowledge. Cambridge: Cambridge University Press (1970).
15. Popper, Karl. Zwei bedeutungen von falsifizierbarkeit. In: Helmut Seiffert, Gerard Radnitzky (eds), Handlexikon zur Wissenschaftstheorie. Munich: Ehrenwirth, 2nd ed., pp. 82–85 (1994).
16. Lakatos, Imre. Falsification and the methodology of research program. In: Imre Lakatos, Alan Musgrave (eds), Criticism and the Growth of Knowledge. Cambridge: Cambridge University Press, pp. 91–197 (1970).
17. Feyerabend, Paul. Against Method. London: New Left Books (1975).
18. Ladyman, J. Understanding Philosophy of Science. London: Routledge (2002).

19. Chalmers, Alan. What Is This Thing Called Science? Indianapolis: Hackett Publishing Company, 4th ed. (2013).
20. Cosmides, L., J. Tooby. Cognitive adaptations for social exchange. In: J. Barkow, L. Cosmides, J. Tooby (eds.), The Adapted Mind. Oxford: Oxford University Press (1992).
21. Tomasello, Michael. Becoming Human: A Theory of Ontogeny. Cambridge, MA: Harvard University Press (2019).
22. Proctor, R., E. Capaldi. Psychology of Science: Implicit and Explicit Processes. Oxford: Oxford University Press (2012).
23. Feist, Gregory, Michael Gorman. Handbook of the Psychology of Science. New York: Springer Press (2013).
24. Heyes, Cecilia. Cognitive Gadgets: The Cultural Evolution of Thinking. Cambridge, MA: Harvard University Press (2018).
25. Stephan, Paula. How Economics Shapes Science. Cambridge, MA: Harvard University Press (2015).
26. Stephan, Paula. The economics of science. Journal of Economic Literature, 34 (3), pp. 1199–1235 (1996).
27. Lane, Julia. Assessing the impact of science funding. Science, 5932 (324), 1273–1275 (2009).
28. Partha, Dasgupta, Paul David. Toward a new economics of science. Research Policy, 23 (5), 487–521 (1994).
29. Greenland, S., S.J. Senn, K.J. Rothman, et al. Statistical tests, P values, confidence intervals, and power: a guide to misinterpretations. European Journal of Epidemiology, 31, 337–350 (2016).
30. Ioannidis, J. Why most published research findings are false. PLoS Medicine, 2 (8) (2005).
31. Hansson, Sven Ove. Science and pseudo-science. Stanford Encyclopedia of Philosophy Archive, https://plato.stanford.edu/archives/sum2017/entries/pseudo-science/ (2017).
32. Kuhn, Thomas, edited by Bojana Mladenovic. The Last Writings of Thomas S. Kuhn: Incommensurability in Science. Chicago: University of Chicago Press (2022).
33. Popper, Karl. Conjectures and Refutations: The Growth of Scientific Knowledge. Melbourne: Hassell Street Press (2021).
34. Popper, Karl. The Logic of Scientific Discovery. London: Routledge, 2nd ed. (1959/2005).
35. Clauset, Aaron, Daniel Larremore, Roberta Sinatra. Data-driven predictions in the science of science. Science, 355, 477–480 (2017).
36. Hu, J. Studying the science of science. Science, March 28 (2016).
37. Sugimoto, Cassidy. Scientific success by numbers. Nature, 593 (2021).
38. Xu, Fengli, Lingfei Wu, James Evans. Flat teams drive scientific innovation. Proceedings of the National Academy of Sciences, 119 (23) e2200927119 (2022).
39. Editorial. Broader scope is key to the future of 'science of science'. Nature Human Behaviour, 6, 899–900 (2022).

40. Ioannidis, J., D. Fanelli, D. Dunne, S. Goodman. Meta-research: evaluation and improvement of research methods and practices. PLoS Biology, 13 (10), e1002264 (2015).
41. Moher, D., S. Hopewell, K. Schulz, et al. CONSORT 2010 Explanation and elaboration: updated guidelines for reporting parallel group randomised trials. BMJ, 340, c869 (2010).
42. Krauss, Alexander. Why all randomised controlled trials produce biased results. Annals of Medicine, 50 (4), 312–322 (2018).
43. Audretsch, David, Barry Bozeman, Kathryn Combs, et al. The economics of science and technology. Journal of Technology Transfer, 27 (2), 155–203 (2002).
44. Gil, Y., M. Greaves, J. Hendler, H. Hirsh. Artificial intelligence. Amplify scientific discovery with artificial intelligence. Science, 346 (6206), 171–172 (2014).
45. Titano, J.J., M. Badgeley, J. Schefflein, et al. Automated deep-neural-network surveillance of cranial images for acute neurologic events. Nature Medicine, 24, 1337–1341 (2018).
46. Alkhateeb, Ahmed, Aeon. Can scientific discovery be automated? The Atlantic, April 25 (2017).
47. Nersessian, Nancy. The cognitive basis of model-based reasoning in science. In: Peter Carruthers, Stephen Stich, Michael Siegal (eds), The Cognitive Basis of Science. Cambridge: Cambridge University Press (2002).
48. Carruthers, Peter, Stephen Stich, Michael Siegal. Introduction: what makes science possible? In: Peter Carruthers, Stephen Stich, Michael Siegal (eds), The Cognitive Basis of Science. Cambridge: Cambridge University Press (2002).
49. Giere, Ronald. Explaining Science: A Cognitive Approach. Chicago: University of Chicago Press (1988).
50. Mercier, Hugo, Dan Sperber. The Enigma of Reason. Cambridge, MA: Harvard University Press (2017).
51. Darwin, Charles. On the Origin of Species by Means of Natural Selection, or the Preservation of Favoured Races in the Struggle for Life. London: John Murray (1859).
52. Laland, Kevin. Darwin's Unfinished Symphony: How Culture Made the Human Mind. Princeton: Princeton University Press (2017).
53. Hull, D. Science and Selection: Essays on Biological Evolution and the Philosophy of Science. Cambridge: Cambridge University Press (2001).
54. Pigliucci, M., J. Kaplan. Making Sense of Evolution: The Conceptual Foundations of Evolutionary Theory. Chicago: University of Chicago Press (2006).
55. Lévi-Strauss, Claude. The Savage Mind. Chicago: University of Chicago Press (1966).
56. Blurton-Jones, N., M. Konner. !Kung knowledge of animal behaviour. In: R. Lee, I. DeVore (eds), Kalahari Hunter–Gatherers. Cambridge: Cambridge University Press (1976).
57. Nisbett, R., K. Peng, I. Choi, A. Norenzayan. Culture and systems of thought: holistic vs. analytic cognition. Psychological Review, 108 (2), 291–310 (2001).
58. Edwards, Jeanette, Penny Harvey, Peter Wade. Anthropology and Science: Epistemologies in Practice. Oxford: Berg Publishers (2007).
59. Hess, David. Social movements and the anthropology of science and technology. American Anthropologist, 109 (3), 463–472 (2007).

60. Mithen, Steven. Human evolution and the cognitive basis of science. In: Peter Carruthers, Stephen Stich, Michael Siegal (eds), The Cognitive Basis of Science. Cambridge: Cambridge University Press (2002).
61. Toth, N., K. Schick, S. Savage-Rumbaugh, R. Sevcik, D. Rumbaugh. Pan the tool maker: investigations into the stone tool-making and tool-using capabilities of a bonobo (*Pan paniscus*). Journal of Archaeological Science, 20 (1993).
62. D'Errico, F. A new model and its implications for the origin of writing: the La Marche antler revisited. Cambridge Archaeological Journal, 5, 81–91 (1995).
63. S. Schneegans, T. Straza, J. Lewis (eds). UNESCO Science Report: The Race Against Time for Smarter Development. Statistical Annex. Paris: UNESCO (2021).
64. Krauss, Alexander. The Motor of Scientific Discovery. Forthcoming (2025).
65. Oxford English Dictionary. Science. https://www.oed.com/dictionary/science_n?tab=factsheet#23959055 (2022).
66. Scopus. Scopus database. Amsterdam: Elsevier, www.scopus.com (2024).
67. Znaniecki, Florian. The subject matter and tasks of the science of knowledge. In: Bohdan Walentynowicz (ed.), Polish Contributions to the Science of Science (translation, 1982). Dordrecht: Reidel Publishing Company (1923).
68. Skalska-Zlat, M. Nalimov and the Polish way towards science of science. Scientometrics, 52, 211–223 (2001).
69. Galton, F. English Men of Science: Their Nature and Nurture. London: MacMillan (1874).
70. de Solla Price, Derek. Little Science, Big Science. New York: Columbia University Press (1963).
71. Zuckerman, H. Scientific Elite: Nobel Laureates in the United States. New York: Free Press (1977).
72. Boersema, Jan. Lucas Reijnders. Principles of Environmental Sciences. Dordrecht: Springer (2009).
73. Thorndike, E. Animal Intelligence. Darien, CT: Hafner Publishing Company (1970).
74. de Waal, Frans. Are We Smart Enough to Know How Smart Animals Are? New York: W.W. Norton & Company (2016).
75. McGrew, W. Chimpanzee Material Culture. Cambridge: Cambridge University Press (1992).
76. Boesch, C., H. Boesch. Possible causes of sex differences in the use of natural hammers by wild chimpanzees. Journal of Human Evolution, 13, 415 (1984).
77. Morell, V. Watch this genius bird plan for its 'handyman' job. Science, February 10 (2023).
78. Herrmann, Esther, Josep Call, Maráa Victoria Hernàndez-Lloreda, Brian Hare, Michael Tomasello. Humans have evolved specialized skills of social cognition: the cultural intelligence hypothesis. Science, 317, 1360 (2007).
79. Garcia, J., B.K. McGowan, F.R. Ervin, R.A. Koelling. Cues: their relative effectiveness as a function of the reinforcer. Science, 160 (3829), 794–795 (1968).
80. Miller, R., S. Owens, B. Rørslett. Plants and colour: flowers and pollination. Optics and Laser Technology, 43, 282–294 (2011).

81. Sullivan, P. Objects limit human comprehension. Biology & Philosophy, 24, 65–79 (2009).
82. Krauss, Alexander. Homo methodologicus and the origin of science and civilisation. Heliyon, 9, 10 (2023).
83. National Science Foundation (NSF). Federal Funds for R&D series. American Association for the Advancement of Science (AAAS), https://www.aaas.org/programs/r-d-budget-and-policy/research-science-and-engineering-discipline (2019).
84. Harmand, Sonia, Jason Lewis, Craig Feibel, et al. 3.3-million-year-old stone tools from Lomekwi 3, West Turkana, Kenya. Nature, 521, 310–315 (2015).
85. Susman, R. Who made the Oldowan tools? Fossil evidence for tool behaviour in Plio-Pleistocene hominids. Journal of Anthropological Research, 47 (2), (1991).
86. Gowlett, J. Mental abilities of early man: a look at some hard evidence. In: R. Foley (ed.), Hominid Evolution and Community Ecology. Cambridge, MA: Academic Press (1984).
87. Sterelny, Kim. The Evolved Apprentice: How Evolution Made Humans Unique. Cambridge, MA: MIT Press (2012).
88. Henrich, Joseph. The Secret of Our Success: How Culture Is Driving Human Evolution, Domesticating Our Species, and Making Us Smarter. Princeton: Princeton University Press (2016).
89. Carruthers, Peter. The roots of scientific reasoning: infancy, modularity and the art of tracking. In: Peter Carruthers, Stephen Stich, Michael Siegal (eds), The Cognitive Basis of Science. Cambridge: Cambridge University Press (2002).
90. Tomasello, Michael. The Cultural Origins of Human Cognition. Cambridge, MA: Harvard University Press (1999).
91. Currie, Adrian. Rock, Bone, and Ruin: An Optimist's Guide to the Historical Sciences. Cambridge, MA: MIT Press (2018).
92. Diamond, J. Guns, Germs, and Steel: The Fates of Human Societies. New York: W.W. Norton & Company (1997).
93. Banning, E. The Neolithic period: triumphs of architecture, agriculture and art. Near Eastern Archaeology, 61 (4), 188–237 (1998).
94. Williams, L. Pearce. History of science. Encyclopaedia Britannica, https://www.britannica.com/science/history-of-science (2017).
95. Freeden, Willi, Clemens Heine, Zuhair Nashed. An Invitation to Geomathematics. Cham: Birkhäuser, Springer Nature (2019).
96. Wilkins, Robert. Neurosurgical Classics. Park Ridge, IL: American Association of Neurological Surgeons, 2nd ed. (1992).
97. Chalmers, I., E. Dukan, S. Podolsky, G. Davey Smith. The advent of fair treatment allocation schedules in clinical trials during the 19th and early 20th centuries. Journal of the Royal Society of Medicine, 105, 221–227 (2012).
98. Peirce, Charles S. Illustrations of the logic of science. Popular Science Monthly, November (1877).
99. Atran, Scott, Douglas Medin. Introduction. In: The Native Mind and the Cultural Construction of Nature. Cambridge, MA: MIT Press (2008).
100. Pinker, S. The cognitive niche: coevolution of intelligence, sociality, and language. Proceedings of the National Academy of Sciences, 107 (2010).

101. Shettleworth, S. Cognition, Evolution and Behaviour. Oxford: Oxford University Press, 2nd ed. (2010).
102. Gigerenzer, Gerd, Henry Brighton. Homo heuristicus: why biased minds make better inferences. Topics in Cognitive Science, 1, 107–143 (2009).
103. Tooby, J., L. Cosmides. The theoretical foundations of evolutionary psychology. In: D.M. Buss (ed.), The Handbook of Evolutionary Psychology, 2nd ed., Vol. 1. New Jersey: Wiley (2016).
104. O'Brien, M., K. Laland. Genes, culture, and agriculture: an example of human niche construction. Current Anthropology, 53 (4), 434–470 (2012).
105. Pinker, S. How the Mind Works. New York: W.W. Norton & Company (2009).
106. Hutmacher, Fabian. Why is there so much more research on vision than on any other sensory modality? Frontiers in Psychology, 10, October 4 (2019).
107. Carnap, Rudolf. Scheinprobleme in der Philosophie (translation, Pseudoproblems in Philosophy). Berlin: Bernary, pp. 301–343 (1928).
108. Bohr, Niels. The quantum postulate and the recent development of atomic theory. Nature, 121, 580–590 (1928).
109. Kahneman, Daniel. Thinking Fast and Slow. New York: Farrar, Straus and Giroux (2011).
110. Tversky, A., D. Kahneman. Judgment under uncertainty: heuristics and biases. Science, 185, 1124–1131 (1974).
111. Zvereva, E.L., M.V. Kozlov. Biases in ecological research: attitudes of scientists and ways of control. Nature Scientific Reports, 11, 226 (2021).
112. Kukucka, Jeff, Saul Kassin, Patricia Zapf, Itiel Dror. Cognitive bias and blindness: a global survey of forensic science examiners. Journal of Applied Research in Memory and Cognition, 6 (4) (2017).
113. O'Sullivan, E.D., S.J. Schofield. A cognitive forcing tool to mitigate cognitive bias—a randomised control trial. BMC Medical Education, 19, 12 (2019).
114. Bishop, Dorothy. The psychology of experimental psychologists: overcoming cognitive constraints to improve research. Quarterly Journal of Experimental Psychology, 73 (1) (2019).
115. Bishop, Dorothy. How scientists can stop fooling themselves over statistics. Nature, 584 (9) (2020).
116. Kerr, N. HARKing: Hypothesizing After the Results are Known. Personality and Social Psychology Review, 2 (3), 196–217 (1998).
117. Rosenthal, R., R. Rosnow. Essentials of Behavioral Research: Methods and Data Analysis. New York: McGraw-Hill, 2nd ed. (1991).
118. Feist, G. Psychology of science as a new subdiscipline in psychology. Current Directions in Psychological Science, 20 (5), 330–334 (2011).
119. Krauss, Alexander. Science's greatest discoverers: a shift towards greater interdisciplinarity, top universities and older age. Nature, Humanities and Social Sciences Communications, 11, 272 (2024).
120. Merton, R. Behavior patterns of scientists. Leonardo, 3 (2), 213–220 (1970).
121. Henry, John. The Scientific Revolution and the Origins of Modern Science. Houndsmill: Palgrave Macmillan, 3rd ed. (2008).
122. Watson, James. The Double Helix. New York: New American Library (1969).

123. Strauss, Mark. Nobel laureates who were not always noble. National Geographic, October 6 (2015).
124. Planck, Max. Scientific Autobiography and Other Papers (translation, Frank Gaynor). New York: Philosophical Library (1949).
125. Usher, I., P. Hellyer, K.S. Lee, et al. "It's not rocket science" and "It's not brain surgery"—"It's a walk in the park": prospective comparative study. BMJ, 375 e067883 (2021).
126. Ericsson, K., R. Hoffman, A. Kozbelt, et al. The Cambridge Handbook of Expertise and Expert Performance. Cambridge: Cambridge University Press (2006).
127. Gascoigne, Robert. The historical demography of the scientific community, 1450–1900. Social Studies of Science, 22 (3), 545–573 (1992).
128. Tweney, R., M. Doherty, C. Mynatt. On Scientific Thinking. New York: Columbia University Press, pp. 325–326 (1981).
129. Harwood, J. Styles of Scientific Thought: The German Genetics Community 1900–1933. Chicago: University of Chicago Press (1993).
130. Merton, Robert. Science and technology in a democratic order. Journal of Legal and Political Sociology, 1, 115–126 (1942).
131. Merton, R. The Matthew effect in science. Science, 159, 56–63 (1968).
132. Sokal, A. Transgressing the boundaries: towards a transformative hermeneutics of quantum gravity. Social Text, 46/47, 217–252 (1996).
133. Gross, P., N. Levitt. Higher Superstition: The Academic Left and Its Quarrels with Science. Baltimore: Johns Hopkins University Press (1994).
134. Encyclopaedia Britannica. Jocelyn Bell Burnell. Encyclopaedia Britannica, https://www.britannica.com/biography/Jocelyn-Bell-Burnell (2023).
135. Reichenbach, H. Experience and Prediction: An Analysis of the Foundations and the Structure of Knowledge. Whitefish, MT: Literary Licensing, LLC (1949).
136. Chen, C. On the shoulders of giants. In: Mapping Scientific Frontiers: The Quest for Knowledge Visualization. London: Springer (2003).
137. Yin, Y., Y. Dong, K. Wang, et al. Public use and public funding of science. Nature Human Behaviour, 6, 1344–1350 (2022).
138. The Nobel Prize. The Nobel Prize. www.nobelprize.org (2024).
139. Krauss, Alexander. External Influences and the Educational Landscape: Analysis of Political, Economic, Geographic, Health and Demographic Factors in Ghana. New York: Springer, pp. 1–143 (2013).
140. Derry, Gregory. What Science Is and How It Works. Princeton: Princeton University Press (2002).
141. Federal Reserve Bank of St. Louis. FRED Economic Data. Investment in government fixed assets. https://fred.stlouisfed.org/release/tables?rid=356&eid=145951#snid=145952 (2020).
142. Yao, L., Y. Li, S. Ghosh, J. Evans, A. Rzhetsky. Health ROI as a measure of misalignment of biomedical needs and resources. Nature Biotechnology, 33, 807–811 (2015).
143. Boudreau, K., T. Brady, I. Ganguli, P. et al. A field experiment on search costs and the formation of scientific collaborations. The Review of Economics and Statistics, 99 (4), 565–576 (2017).

144. Bol, T., M. de Vaan, A. van de Rijt. The Matthew effect in science funding. Proceedings of the National Academy of Sciences, 115 (19), 4887–4890 (2018).
145. Azoulay, Pierre, Christian Fons-Rosen, Joshua S. Graff Zivin. Does science advance one funeral at a time? American Economic Review, 109 (8), 2889–2920 (2019).
146. Walsh, E., M. Rooney, L. Appleby, G. Wilkinson. Open peer review: a randomised controlled trial. British Journal of Psychiatry, 176 (1), 47–51 (2000).
147. Kirk, Geoffrey, John Raven, Malcolm Schofield. The Presocratic Philosophers: A Critical History with a Selection of Texts. Cambridge: Cambridge University Press, 2nd ed. (1983).
148. Zirkle, Conway. Natural selection before the 'Origin of Species'. Proceedings of the American Philosophical Society, 84 (1), 71–123 (1941).
149. Bowler, Peter. Evolution: The History of an Idea. Berkeley: University of California Press, 3rd ed. (2003).
150. Darwin, Erasmus. Zoonomia. New York: AMS Press (1794).
151. Keynes, Richard. Charles Darwin's Zoology Notes and Specimen Lists from H.M.S. Beagle. Cambridge: Cambridge University Press (2000).
152. Boughn, Stephen. Hasenöhrl and the equivalence of mass and energy. European Physical Journal H (2013).
153. Bartocci, Umberto. Albert Einstein e Olinto De Pretto. La Vera Storia della Formula Più Famosa del Mondo (translation, Albert Einstein and Olinto De Pretto. The True History of the Most Famous Formula in the World). Bologna: Andromeda (1999).
154. Janssen, Michel, Matthew Mecklenburg. From classical to relativistic mechanics: electromagnetic models of the electron. In: Vincent F. Hendricks, Klaus Frovin Jørgensen, Jesper Lützen, Stig Andur Pedersen (eds), Interactions: Mathematics, Physics and Philosophy, 1860–1930. Dordrecht: Springer (2007).
155. Einstein, Albert. Das prinzip von der erhaltung der schwerpunktsbewegung und die trägheit der energie. Annalen der Physik, 20 (8), 627–633 (1906).
156. Einstein, Albert. Über die vom relativitätsprinzip geforderte trägheit der energie (translation, On the inertia of energy required by the relativity principle). Annalen der Physik, 23, 371–384 (1907).
157. Needham, Joseph. The Grand Titration: Science and Society in East and West. Toronto: University of Toronto Press (1969).
158. Logan, R. The Alphabet Effect. New York: Morrow (1986).
159. Needham, Joseph. Science and Civilisation in China. Cambridge: Cambridge University Press, Vol. 3, pp. 154, 167–168 (1956).
160. Needham, Joseph. Foreword. In: D. Raven, W. Krohn, R.S. Cohen (eds), Edgar Zilsel, The Social Origins of Modern Science. Dordrecht: Kluwer Academic Publishers (2003).
161. Sheldrake, Rupert. Science Set Free: 10 Paths to New Discovery. New York City: Harmony/Rodale (2012).
162. Baker, M. 1,500 scientists lift the lid on reproducibility. Nature, 533, May 26 (2016).
163. Shi, F., J. Foster, J. Evans. Weaving the fabric of science: dynamic network models of science's unfolding structure. Social Networks, 43, 73–85 (2015).

164. Li, J., Y. Yin, S. Fortunato, D. Wang. Scientific elite revisited: patterns of productivity, collaboration, authorship and impact. Journal of the Royal Society of Interface, 17, 20200135 (2020).
165. Rzhetsky, A., J. Foster, I. Foster, J. Evans. Choosing experiments to accelerate collective discovery. Proceedings of the National Academy of Sciences, 112, 14569–14574 (2015).
166. Uzzi, B., S. Mukherjee, M. Stringer, B. Jones. Atypical combinations and scientific impact. Science, 342, 468–472 (2013).
167. Shi, F., J. Evans. Surprising combinations of research contents and contexts are related to impact and emerge with scientific outsiders from distant disciplines. Nature Communications, 14, 1641 (2023).
168. Azoulay, P., J. Graff Zivin, G. Manso. Incentives and creativity: evidence from the academic life sciences. The RAND Journal of Economics, 42, 527–554 (2011).
169. Jones, B., E. Reedy, B. Weinberg. Age and scientific genius. In: D.K. Simonton (ed.), The Wiley Handbook of Genius. Chichester: John Wiley & Sons, pp. 422–450 (2014).
170. Simonton, D. Creative productivity: a predictive and explanatory model of career trajectories and landmarks. Psychological Review, 104, 66–89 (1997).
171. Jones, Benjamin. The burden of knowledge and the "death of the Renaissance Man": is innovation getting harder? The Review of Economic Studies, 76, 1, 283–317 (2009).
172. Wuchty, S., B. Jones, B. Uzzi. The increasing dominance of teams in production of knowledge. Science, 316, 1036–1039 (2007).
173. Wu, L., D. Wang, J.A. Evans. Large teams develop and small teams disrupt science and technology. Nature, 566, 378–382 (2019).
174. Danús, Lluís, Carles Muntaner, Alexander Krauss, Marta Sales-Pardo, Roger Guimerà. Differences in collaboration structures and impact among prominent researchers in Europe and North America. EPJ Data Science, 12 (2023).
175. Krauss, Alexander, Lluís Danús, Marta Sales-Pardo. Early-career factors largely determine the future impact of prominent researchers: evidence across eight scientific fields. Nature, Scientific Reports, 13, 18794 (2023).
176. Li, W., T. Aste, F. Caccioli, et al. Early coauthorship with top scientists predicts success in academic careers. Nature Communications, 10, 5170 (2019).
177. Clarivate. History of citation indexing. https://clarivate.com/webofsciencegroup/essays/history-of-citation-indexing/ (2024).
178. Martin, A., R. Synge. A new form of chromatogram employing two liquid phases. Biochemistry Journal, 35 (12), 1358–1368 (1941).
179. Gordon, J., H. Zeiger, C. Townes. Molecular microwave oscillator and new hyperfine structure in the microwave spectrum of NH_3. Physical Review, 95, 282 (1954).
180. Knoll, M., E. Ruska. Das elektronenmikroskop. Zeitschrift für Physik, 78, 318–339 (1932).
181. Svedberg, The, Herman Rinde. The ultra-centrifuge, a new instrument for the determination of size and distribution of size of particle in amicroscopic colloids. Journal of the American Chemical Society, 46 (12), 2677–2693 (1924).
182. Britannica, The Editors of Encyclopaedia. "Yongle dadian." *Encyclopedia Britannica*, https://www.britannica.com/topic/Yongle-dadian (2015).

183. Shannon, Claude. A mathematical theory of communication. The Bell System Technical Journal, 27, 379–423, 623–656 (1948).
184. Haven, Kendall. 100 Greatest Science Discoveries of All Time. London: Libraries Unlimited (2007).
185. Krenn, M., R. Pollice, S.Y. Guo, et al. On scientific understanding with artificial intelligence. Nature Reviews Physics, 4, 761–769 (2022).
186. Hollon, T.C., B. Pandian, A.R. Adapa, et al. Near real-time intraoperative brain tumor diagnosis using stimulated Raman histology and deep neural networks. Nature Medicine, 26, 52–58 (2020).
187. The Royal Society. The AI Revolution in Scientific Research. London: The Royal Society and The Alan Turing Institute (2019).
188. Langley, P., A. Arvay. Scientific discovery, process models, and the social sciences. In: M. Addis, P. Lane, P. Sozou, F.Gobet (eds) Scientific Discovery in the Social Sciences. Cham (Switzerland): Springer (2019).
189. Schmidt, M., H. Lipson. Distilling free-form natural laws from experimental data. Science, 324, 81–85 (2009).
190. Džeroski, S., L. Todorovski. Equation discovery for systems biology: finding the structure and dynamics of biological networks from time course data. Current Opinion in Biotechnology, 19, 360–368 (2008).
191. Simon, H. A., Langley, P., & Bradshaw, G. L. Scientific discovery as problem solving. Synthese, 47, 1–27 (1981).
192. Khoury, M., J. Ioannidis. Big data meets public health. Science, 346, 6213 (2014).
193. Fisher, R.A. Statistical Methods for Research Workers. Edinburgh: Oliver & Boyd (1925).
194. Ioannidis, J., G. Witte. Reproducibility in crisis: sample quality and the importance of early and ongoing analysis. Science, March 14 (2018).
195. Munafò, M., B. Nosek, D. Bishop, et al. A manifesto for reproducible science. Nature Human Behaviour, 1 (0021) (2017).
196. Allison, D., Andrew W. Brown, Brandon J. George, Kathryn A. Kaiser. Reproducibility: a tragedy of errors. Nature, 530, 27–29 (2016).
197. Goldacre, B. Make journals report clinical trials properly. Nature. 530, 7 (2016).
198. Krauss, Alexander. Multi-studies: a novel approach to addressing irreplicability in RCTs. In: Margaret MacDougall (ed.). A Medical Educator's Guide to Thinking Critically about Randomised Controlled Trials: Deconstructing the "Gold Standard". Cham (Switzerland): Springer (2024).
199. Frenken, Koen. Geography of Scientific Knowledge: A Proximity Approach. Working Papers. Eindhoven: Eindhoven Center for Innovation Studies (2010).
200. Leonelli, Sabina. Data-Centric Biology: A Philosophical Study. Chicago: University of Chicago Press (2016).
201. Cartwright, N. What are randomised controlled trials good for? Philosophical Studies, 147, 59 (2010).
202. Peters, Uwe, Alexander Krauss, Oliver Braganza. Generalization bias in science. Cognitive Science, 46, e13188 (2022).
203. Godfrey-Smith, P. Theory and Reality: An Introduction to the Philosophy of Science. Chicago: University of Chicago Press (2003).

204. Okasha, Samir. Philosophy of Science. A Very Short Introduction. Oxford: Oxford University Press (2002).
205. Curd, M., J. Cover, C. Pincock. Philosophy of Science: The Central Issues. New York: W.W. Norton & Company, 2nd ed. (2012).
206. Hume, David. A Treatise of Human Nature. London: John Noon (1739).
207. Schurz, Gerhard. Hume's Problem Solved: The Optimality of Meta-Induction. Cambridge, MA: MIT Press (2019).
208. Kuhn, Thomas. Logic of discovery or psychology of research? In: P.A. Schilpp (ed.), The Philosophy of Karl Popper. The Library of Living Philosophers, Volume xiv, Book II. La Salle: Open Court, pp. 798–819 (1974).
209. van Fraassen, Bas. The Scientific Image. Oxford: Oxford University Press (1980).
210. van Fraassen, Bas. Scientific Representation: Paradoxes of Perspective. Oxford: Oxford University Press (2008).
211. Kant, Immanuel. Critique of Pure Reason (translation, 1929). London: MacMillan (1781).
212. Plato. The Theaetetus of Plato, with a revised text and English notes by Louis Campbell. Oxford: Oxford University Press (1861).
213. Wittgenstein, Ludwig. Tractatus Logico-Philosophicus. London: Kegan Paul (1922).
214. Fowler, Thomas (ed.). Bacon's Novum Organum. Oxford: Clarendon Press (1878).
215. Thornton, Stephen. Karl Popper. Stanford Encyclopedia of Philosophy Archive, https://plato.stanford.edu/archives/sum2017/entries/popper/ (2017).
216. Mumford, Stephen, Matthew Tugby (eds). Metaphysics and Science. Oxford: Oxford University Press (2013).
217. Dilworth, Craig. The Metaphysics of Science: An Account of Modern Science in Terms of Principles, Laws and Theories. Dordrecht: Springer (2006).
218. Schrenk, Markus. Metaphysics of Science: A Systematic and Historical Introduction. London: Routledge (2017).
219. Earman, John, Clark Glymour, Sandra Mitchell (eds). Ceteris Paribus Laws. Springer, Erkenntnis (2002). https://link.springer.com/article/10.1023/A:1021536426130.
220. Cartwright, Nancy. The Dappled World: A Study of the Boundaries of Science. New York: Cambridge University Press (1999).
221. Donahue, William. New Astronomy (1609) (translation, Astronomia Nova). Cambridge: Cambridge University Press (1992).
222. Stillman, Drake. The Assayer (1623) by Galileo Galilei (translation, Il Saggiatore). Philadelphia: University of Pennsylvania Press (1960).
223. Newton, Isaac. Philosophiæ Naturalis Principia Mathematica. Pepys: London (in English 1728) (1687).
224. Einstein, Albert. Does inertia of a body depend on its energy content? (translation). Annalen der Physik, 18, 639–641 (1905).
225. Encyclopaedia Britannica. Scientific method. https://www.britannica.com/science/scientific-method (2022).
226. Collins Dictionary. Science. www.collinsdictionary.com/dictionary/english/science (2022).
227. Cambridge Dictionary. Science. https://dictionary.cambridge.org/dictionary/english/science (2022).

228. Encyclopaedia Britannica. Science. https://www.britannica.com/science/science (2022).
229. Sober, Elliott. Ockham's Razors: A User's Manual. Cambridge: Cambridge University Press (2015).
230. Weinberg, S. Toward the final laws of physics. In: R. Feynman, S. Weinberg (eds), Elementary Particles and the Laws of Physics. Cambridge: Cambridge University Press (1987).
231. Feynman, R.P., R.B. Leighton, M. Sands. The Feynman Lectures on Physics. Reading, MA: Addison-Wesley (1963).
232. Hawking, S. Is the End in Sight for Theoretical Physics? Cambridge: Cambridge University Press (1988).
233. National Academy of Sciences. Science, Evolution, and Creationism. Washington, DC: The National Academies Press (2008).
234. Einstein, Albert. The World As I See It. London: John Lane, p. 29 (1935).
235. Larivière, V., S. Haustein, K. Börner. Long-distance interdisciplinarity leads to higher scientific impact. PLoS ONE, 10, e0122565 (2015).
236. Gintis, H. A framework for the unification of the behavioral sciences. Behavioral and Brain Sciences, 30 (1), 1–61 (2007).
237. Merriam-Webster Dictionary. Scientific method. www.merriam-webster.com/dictionary/scientific_method (2022).
238. Krauss, Alexander. Redefining the scientific method: as the use of sophisticated scientific methods that extend our mind. PNAS Nexus, 3 (4), 112 (2024).
239. Chang, Hasok. Inventing Temperature: Measurement and Scientific Progress. Oxford: Oxford University Press, p. 238 (2004).
240. Kaplan, Abraham. The Conduct of Inquiry: Methodology for Behavioral Science. San Francisco: Chandler Publishing Company (1964).
241. Spelke, E., S. Tsivkin. Language and number: a bilingual training study. Cognition, 78, 45–88 (2001).
242. Bucchi, Massimiano, Brian Trench. Routledge Handbook of Public Communication of Science and Technology. London: Taylor and Francis, 2nd ed. (2014).
243. Gustafson, Abel, Ronald Rice. The effects of uncertainty frames in three science communication topics. Science Communication, 41 (6), 679–706 (2019).
244. Dunbar, K., L.A. Baker. Goals, analogy, and the social constraints of scientific discovery. Behavioral and Brain Sciences, 17, 538–539 (1994).
245. The Nobel Prize. The Nobel Prize in Chemistry 2020. https://www.nobelprize.org/prizes/chemistry/2020/press-release/ (2020).
246. LIGO Laboratory. Facts. Pasadena: California Institute of Technology, https://www.ligo.caltech.edu/page/facts (2022).
247. Kennedy, D., C. Norman (eds). What don't we know? Science, July 1, 78, 309 (2005).
248. Gleiser, Marcelo. The Island of Knowledge: The Limits of Science and the Search for Meaning. New York: Basic Books (2014).
249. du Sautoy, Marcus. What We Cannot Know: Explorations at the Edge of Knowledge. London: 4th Estate (2016).
250. Barrow, John. Impossibility: The Limits of Science and the Science of Limits. Oxford: Oxford University Press (1999).

251. Dewdney, A.K. Beyond Reason: 8 Great Problems That Reveal the Limits of Science. Hoboken, NJ: Wiley (2004).
252. Yanofsky, Noson. The Outer Limits of Reason: What Science, Mathematics, and Logic Cannot Tell Us. Cambridge, MA: MIT Press (2013).
253. Horgan, John. The End of Science: Facing the Limits of Knowledge in the Twilight of the Scientific Age. New York: Basic Books (2015).
254. Fodor, J. The Modularity of Mind: An Essay on Faculty Psychology. Cambridge, MA: MIT Press (1983).
255. Chomsky, N. New Horizons in the Study of Language and Mind. Cambridge: Cambridge University Press (2000).
256. Chomsky, N. Science, mind, and limits of understanding. In: A.J. Gallego, R. Martin (eds), Language, Syntax, and the Natural Sciences. Cambridge: Cambridge University Press (2018).
257. Williamson, Timothy. Knowledge and Its Limits. Oxford: Oxford University Press (2002).
258. Rescher, Nicholas. The Limits of Science. Pittsburgh: University of Pittsburgh Press (1999).
259. Pigliucci, Massimo, Maarten Boudry (eds). Philosophy of Pseudoscience: Reconsidering the Demarcation Problem. Chicago: University Chicago Press (2013).
260. Bardell, David. The invention of the microscope. Bios, 75 (2), 78–84 (2004).
261. The Nobel Prize. The Nobel Prize in Physics 1986. https://www.nobelprize.org/prizes/physics/1986/press-release/ (1986).
262. Gribbin, John. Britannica Guide to 100 Most Influential Scientists. London: Robinson Publishing (2008).
263. Popper, Karl. The Logic of Scientific Discovery. New York: Harper (1959).
264. Lakatos, Imre, edited by John Worrall and Gregory Currie. The Methodology of Scientific Research Programmes: Philosophical Papers. Cambridge: Cambridge University Press (1978).
265. Macmillan Dictionary. Theory. www.macmillandictionary.com/dictionary/british/theory (2022).
266. Oxford English Dictionary. Theory. https://www.oed.com/dictionary/theory_n?tl=true (2022).
267. Collins Dictionary. Theory. www.collinsdictionary.com/dictionary/english/theory (2022).
268. Dawkins, Richard. The Greatest Show on Earth: The Evidence for Evolution. New York: Free Press (2009).
269. Hoyninengen-Huene, Paul. Systematicity: The Nature of Science. Oxford: Oxford University Press (2013).
270. The Nobel Prize. The Nobel Prize in Physics 2020. https://www.nobelprize.org/prizes/physics/2020/press-release/ (2020).
271. Weeks, Mary Elvira. The discovery of the elements. XIII. Some spectroscopic discoveries. Journal of Chemical Education, 9 (8) (1932).
272. Emwas, A.H.M., R.M. Salek, J.L. Griffin, et al. NMR-based metabolomics in human disease diagnosis: applications, limitations, and recommendations. Metabolomics, 9, 1048–1072 (2013).

273. Del Sole, R., P. Chiaradia. Surfaces, optical properties of. Encyclopedia of Condensed Matter Physics, pp. 144–150 (2005).
274. Robson, Annie-Louise, Paul C. Dastoor, Jamie Flynn, et al. Advantages and limitations of current imaging techniques for characterizing liposome morphology. Frontiers in Pharmacology, 9 (80) (2018).
275. Sinitsyn, Nikolai. Computing with a single qubit faster than the computation quantum speed limit. Physics Letters A, 382 (7), 477–481 (2018).
276. Fardouly, J., R.D. Crosby, S. Sukunesan. Potential benefits and limitations of machine learning in the field of eating disorders: current research and future directions. Journal of Eating Disorders, 10, 66 (2022).
277. Greene, Brian. String theory. Encyclopaedia Britannica, https://www.britannica.com/science/string-theory (2022).
278. Frodeman, Robert, Julie Thompson Klein, Roberto Carlos Dos Santos Pacheco (eds). The Oxford Handbook of Interdisciplinarity. Oxford: Oxford University Press, 2nd ed. (2017).

Acknowledgements

I am thankful for comments from Corinna Peters, Uwe Peters, Michael Stuart, Nikolas Schöll, Samuli Reijula, Martin Zach, J.P. Grodniewicz, Christopher Evans and Dan Taber. I am also grateful for funding received by the European Commission (Marie Curie programme) and the Ministry of Science and Innovation of the Government of Spain (grant RYC2020-029424-I and PID2021-126200NB-I00).

Index

abstraction 28, 139
age 75
Anthropology of science 60–65
Archaeology of science 27–32
artificial intelligence 80–81

biases, psychological 38–39
biases, statistical 86–87, 89, 147–148
Biology of science 23–26

causal inference 31, 51–52, 95–96, 166–167
citations 4, 74, 76–77
civilization 29–31
Cognitive science of science 33–37
collaborations 51, 75
Computer science of science 79–82

discoveries 67–69, 75, 107–109, 122–123, 146

Economics of science 48–52
experimentation 31, 84

falsification 92–93
funding 25–26, 48–51
future science 155–160

gender 45–46

History of science 53–59
Homo sapiens 28–29

impact 74–77
induction 92
institutions 48
instruments 7–11, 13–14, 35, 49–50, 66–73, 77, 104–113, 114–117, 120–126, 128–143, 145–160, 161–163

Kuhn, Thomas 3, 53–54, 55–56

limits of science 118–160
Linguistics of science 101–103

Matthew effect 43, 51
mind 33–35
Merton, Robert 39–40, 43
metaphysics 93–97
methodological abilities 7, 27, 31, 33, 35, 66, 134–135, 138–139
Methodological toolbox 7–10, 19*f*, 66, 108*f*, 125*f*, 130–134, 136*t*, 140, 161–162
Methodological tower of science 7–11
Methodology of science 66–73
methods 7–11, 13–14, 35, 49–50, 66–73, 77, 104–113, 114–117, 120–126, 128–143, 145–160, 161–163

New-methods-drive-science theory 114–116, 162–163
norms 41–42

observation 35–37, 63–64, 91–92
OVER criterion of science 128–131, 135–138, 140–142

paradigm shifts 3, 53–54
personality traits 39–40
Philosophy of science 91–100
Popper, Karl 3–4, 92–93
Psychology of science 38–40
publications 16–17

randomised controlled trials (RCT) 88, 95, 117, 147–148, 151*t*
replication 86–87, 89
reward system 48

science 13–14
science of science 1–168
scientific method 95–96
Scientometrics and network science 4, 74–78
Sociology of science 41–47
Statistics and mathematics of science 83–90

teams 75
tools *see* instruments

Zuckerman, Harriet 42–43